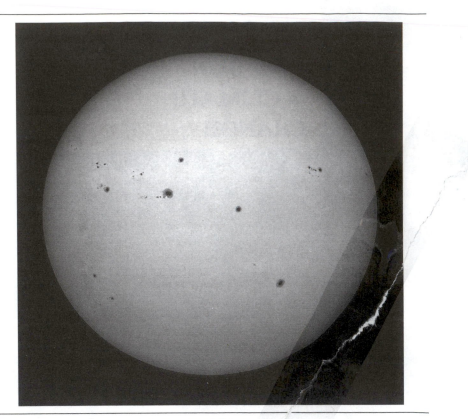

THE SUN ON JULY 6, 1979. FROM W. J. LIVINGSTON.

The ROLE
of the SUN
in CLIMATE
CHANGE

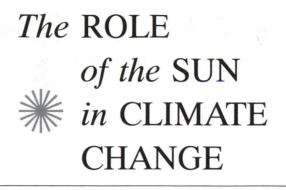

Douglas V. Hoyt

Kenneth H. Schatten

New York   Oxford • Oxford University Press   1997

Oxford University Press

Oxford   New York
Athens   Auckland   Bangkok   Bogota   Bombay   Buenos Aires
Calcutta   Cape Town   Dar es Salaam   Delhi   Florence   Hong Kong
Istanbul   Karachi   Kuala Lumpur   Madras   Madrid   Melbourne
Mexico City   Nairobi   Paris   Singapore   Taipei   Tokyo   Toronto

and associated companies in
Berlin   Ibadan

Published by Oxford University Press, Inc.,
198 Madison Avenue, New York, New York 10016

Library of Congress Cataloging-in-Publication Data
Hoyt, Douglas V.
The role of the sun in climate change / Douglas V. Hoyt, Kenneth H. Schatten.
    p.   cm.
Includes bibliographical references and index.
ISBN 0-19-509413-1; ISBN 0-19-509414-X (pbk.)
    1. Solar activity.   2. Climatic changes.   I. Schatten, Kenneth H.   II. Title.
QC883.2.S6H69   1997
551.6—dc20   96-10848

9 8 7 6 5 4 3 2 1

Printed in the United States of America
on acid-free paper

Acknowledgments

We would like to thank Tom Bryant, Richard A. Goldberg, and O. R. White for reviewing a draft of this book. Their comments helped improve the book. Dr. Elena Gavryuseva and Dr. Ron Gilliland sent us the neutrino-flux calculations. Dr. Eugene Parker gave us an estimate of the energy-storage requirements in the solar convection zone associated with long-term changes in solar luminosity. Ruth Freitag of the Library of Congress aided in tracking down some biographical information. Any errors are solely the responsibility of the authors, and any views expressed here do not reflect any organizational viewpoints. Finally, one reviewer of this book, who wishes to remain anonymous, receives our heartfelt thanks for greatly improving the readability of the text.

This book is dedicated to all the pioneers of sun/climate studies.

✳

Contents

The ROLE
of the SUN
in CLIMATE
CHANGE

1. *Introduction*

About 400 years before the birth of Christ, near Mt. Lyscabettus in ancient Greece, the pale orb of the sun rose through the mists. According to habit, Meton recorded the sun's location on the horizon. In this era when much remained to be discovered, Meton hoped to find predictable changes in the locations of sunrise and moonrise. Although rainy weather had limited his recent observations, this foggy morning he discerned specks on the face of the sun, the culmination of many such blemishes in recent years. On a hunch, Meton began examining his more than 20 years of solar records. These seemed to confirm his belief: when the sun has spots, the weather tends to be wetter and rainier.

Theophrastus reported these findings in the fourth century B.C. Other ancient accounts concerning the sun and weather are vague. If one stretches one's imagination, some comments by Aratus of Soli, Virgil, and Pliny the Elder may touch on this subject. What happened to the original records used by Theophrastus? Perhaps these and related scientific data were burned in the fire that destroyed the Library at Alexandria around A.D. 300. Other possible ancient accounts have vanished.

Two thousand years passed. The Roman Empire rose and fell, the Dark Ages lasted a thousand years, and Europe entered the Renaissance. The 1600s reveal perhaps half a dozen scattered references to changes in the sun and their effect on weather. After a few more references in the 1700s, scientific interest in the sun waned. Following Sir William Herschel's comments on sunspots and climate in 1796 and 1801, about 10 scientific papers touched on the sun's influence on climate and weather. The next two decades contain about 10 or so references to this topic. Shortly after a paper by C. Piazzi Smyth appeared in the proceedings of the Royal Society in 1870, the field exploded. This paper stimulated scientists such as Sir Norman Lockyer, Ferguson, Meldrum, and others to think about solar and terrestrial changes. Meldrum, a British meteorolo-

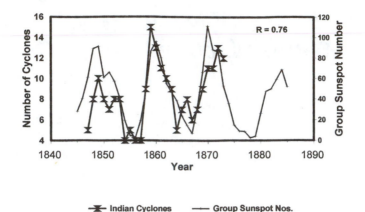

➤ Indian Cyclones   —— Group Sunspot Nos.

FIGURE 1.1 Indian Ocean cyclones and group sunspot numbers. One of the first published claims concerning a relationship between solar activity and terrestrial weather, Dr. Meldrum's data for the number of Indian cyclones from 1847 to 1873 are plotted versus sunspot numbers. This striking relationship inspired many follow-up studies, as well as the first wave of sun/climate investigations (see Chapter 7). (Data for original figure comes from Meldrum 1872, 1885.)

gist in India, considered Indian cyclones. His tabular values are compared with sunspot numbers in Figure 1.1.

The obvious and striking parallelism between the two curves convinced many scientists of the reality of the sun/climate relationship, and investigations began in earnest. Over the next two decades, dozens of papers appeared relating changes in the sun to variations in the Earth's temperature, rainfall and droughts, river flow, cyclones, insect populations, shipwrecks, economic activity, wheat prices, wine vintages, and many other topics. Although many independent studies reached similar conclusions, some produced diametrically opposed results. Certain studies were criticized as careless. Questions critics asked included: Why were people getting different answers at different locations? Why did some relationships exist for an interval and then disappear? Were all these results mere coincidences? Often, "persistence" and "periodicities" in two parallel time series can create the appearance of a coincidental relationship. These statistical problems are covered in chapter 5.

To complicate the issue further, some scientists believed that the sun's variations could explain everything about weather and climate. Other critics countered that the reverse was true, and by the late 1890s the initial enthusiasm concerning the sun and its potential effects on the weather had waned to such an extent that few publications can be found. The critics appeared victorious, and the field nearly died. After this brief hiatus, a steady increase in the number of sun/climate studies has appeared in the twentieth century. Unfortunately, none of these new studies is definitive in either proving or disproving the sun/climate connection.

Before writing this book, we compiled a bibliography of nearly 2,000 papers and books concerning the sun's influence on weather and climate. Figure 1.2 shows the number of publications per year. Although incomplete (no doubt some technical reports and popular accounts were either missed or purposely omitted), our bibliography may be the most comprehensive assemblage of significant papers to date. To our knowledge, thus far no one has read all 20,000-plus pages of text in at least a dozen languages. Furthermore, many papers demonstrate poor statistical analyses, are too enthusiastic in their conclusions, or are repetitive. Critics today might even categorize these papers as fringe science and suggest they be ignored. Indeed, they might characterize the whole field as "pathological science." Whether this harsh judgment is justified remains to be seen. Although many scientists have arrived at the same conclusions while remaining entirely unaware of their colleagues' work, many reported effects are associated with incorrect or inadequate statistics. Rather than being a repository of absolute truths, the scientific literature remains an ongoing debate and discussion. Some erroneous conclusions are always published; however, such errors should not invalidate an entire field of study.

Rather than reviewing innumerable papers, we approach sun/climate change as one might an ongoing journey, highlighting only the better studies and those intriguing results we consider scientifically interesting. Our book is divided into three parts.

1. We start with an examination of solar activity and travel through history to reveal the slow development of our understanding of the sun. Observational accounts will be followed by a description of present-day solar theories. We will then examine why the sun varies and place the sun's variation within the context of other stars.

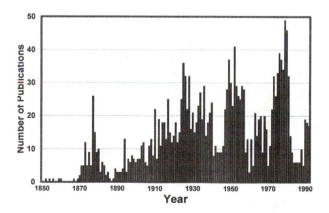

FIGURE 1.2 The approximate number of sun/weather/climate publications each year from 1850 to 1992 are shown (1,908 total). Note the initial surge of publications after 1870 followed by a decline around 1900. Since then, the increase in publications has remained almost steady. Two thousand papers represent less than 0.25% of the scientific literature published *each year*, so the sun/climate field remains relatively small.

2. The central portion of this volume considers climate and the sun/climate connection, particularly on the 11-year time scale. We define what climate is and how sensitive climate would be to changes in the sun's radiative output. We examine how difficult it is to make consistent weather observations over many years; even with good climatic measurements, the weather proves so variable that a solar influence can only be detected on large spatial scales over long intervals. We consider the problem of sampling and its influence on our studies. In addition, we look at the theoretical framework for climate and climatic change. We review the possible sensitivity of Earth's climate to solar changes and advance a new hypothesis that may explain why climate appears more sensitive to solar changes than is generally thought. We can then explore the statistical sun/climate relationships from an informed viewpoint. Four chapters are devoted to studies of temperature, rainfall, storms, and biota, generally proceeding from those results that many scientists would agree warrant consideration, if not further study, to those ideas that initially seem wild and strange. We round out this second part of the book with a discussion of cyclomania, or the search for cycles in the climate and the sun.

3. Finally, we discuss possible alternative explanations for variations in the sun and climate on time scales from decades to billions of years. These solar variations seem to parallel modern reconstruction of climate variations remarkably well. As for decades to centuries, convincing arguments can be developed that the sun is a driving force behind climatic change. To place the solar connection within the context of other ideas, we examine various competing climate theories and explain how climatic change may be deduced by combining several theories. We explore the problem of the early faint sun and the paradox that climate has remained stable for billions of years despite a dramatic increase in the sun's brightness. We summarize several ideas that might account for this paradox, paying particular attention to the Athenian Hypothesis and the popular Gaia Hypothesis.

A concluding chapter details some ironies, as well as arguments, both pro and con, in the field of sun/climate connections. The question of sun/climate connections remains controversial and volatile, and only more experimental and theoretical work will lead to the truth. Throughout the book, we will be presenting evidence on both sides of the question "Does the sun affect the climate?" This may appear confusing to some; however, scientists reach conclusions by examining both sides of an issue, and then seeing which is better justified.

The book has three appendices. Appendix 1 is a glossary of solar and terrestrial terms and their definitions. Appendix 2 tabulates some useful facts and numbers associated with the sun. Appendix 3 provides a technical description of some of the statistical techniques used in many sun/climate and sun/weather studies. The bibliography of sun and climate concludes the book. References to publications in the text are generally mentioned informally, but are listed chapter by chapter. Also included here is a general reference list of early and important books and papers.

I. THE SUN

2. *Observations of the Sun*

A Modern Overview of the Sun

Our sun is a typical "second generation," or G2, star nearly 4.5 billion years old. The sun is composed of 92.1% hydrogen and 7.8% helium gas, as well as 0.1% of such all-important heavy elements as oxygen, carbon, nitrogen, silicon, magnesium, neon, iron, sulfur, and so forth in decreasing amounts (see Appendix 3). The heavy elements are generated from nucleosynthetic processes in stars, novae, and supernovae after the original formation of the Universe. This has led to the popular statement that we are, literally, the "children of the stars" because our bodies are composed of the elements formed inside stars.

From astronomical studies of stellar structure, we know that, since its beginnings, the sun's luminosity has gradually increased by about 30%. This startling conclusion has raised the so-called faint young sun climate problem: if the sun were even a few percent fainter in the past, then Earth could have been covered by ice. In this frozen state, it might not have warmed because the ice would reflect most of the incoming solar radiation back into space. Although volcanic aerosols covering the ice, early oceans moderating the climate, and other theories have been suggested to circumvent the "faint young sun" problem, how Earth escaped the ice catastrophe remains uncertain.

How can the sun generate vast amounts of energy for billions of years and still keep shining? Before nuclear physics, scientists believed the sun generated energy by means of slow gravitational collapse. Still, this process would only let the sun shine about 30 million years before its energy was depleted. To shine longer, the sun requires another energy source. We now believe that a chain of nuclear reactions occurs inside the sun, with four hydrogen nuclei fusing into one helium nucleus at the sun's center. Because the four hydrogen nuclei have more mass than the one helium nucleus, the resulting mass deficit is converted into energy according to Einstein's famous formula $E = mc^2$.

9

The energy, produced near the sun's center, creates a central temperature of about 15 million degrees Kelvin (°K). This same energy is transported from the interior first by radiation and then by convection in the outer layers, ultimately leading to the energy deposition in the surface layers (the photosphere) at 5780 °K. Here the energy is finally radiated into space, and a small fraction bathes our planet with heat and light. Figure 2.1 shows a schematic cross-section of the sun's internal structure.

Dynamo processes in the sun's outer layers, or convection zone, create a magnetic field. This results in sunspots, flares, coronal mass ejections, and other types of "magnetic activity," as well as "the solar cycle." Solar cycles are the periodic variations of the sun's activity and inactivity, varying within an 11-

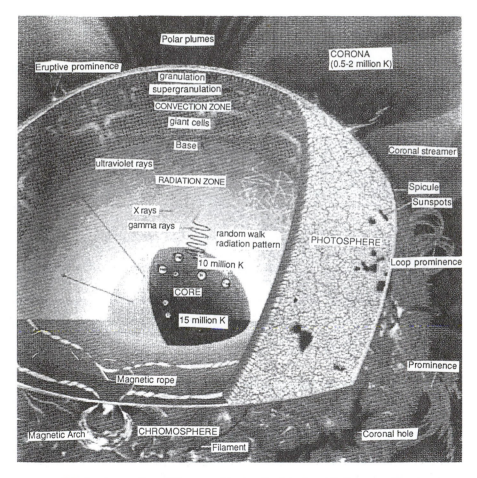

FIGURE 2.1 A cross-section of the sun, showing the interior radiative core, the convective envelope, the photosphere, and surrounding corona. (Adapted from Friedman, 1986, with permission of the author.)

year period. Along with the 11-year variations are longer duration changes such as the "Gleissberg" cycle with time-scale variations of approximately 100 years. These long-period solar variations make the sun a unique candidate for influencing our climate over extended time scales. Other terrestrial variations (e.g., volcanic aerosols) may influence climate for a few years, but might not "drive" the climate system with the long-time-scale forcing needed to provide anything beyond irregular, temporary disturbances.

Sunspots are part of solar "active regions" famous for their flares, coronal mass ejections, and other forms of activity. These features result when the sun's surface magnetic field gains sufficient strength to inhibit the convective heat flow from the sun's interior. Because sunspots are 1500 °K cooler than the sun's surface, when sunspot activity is centrally located on the solar disk (the sun's rotation period is about 27 days), the sun's energy radiated toward Earth is reduced. Space satellites have observed this approximately 0.1% energy reduction, which by itself is probably not sufficient to influence climate. The average energy radiated to Earth, known as the sun's total irradiance or "solar constant," was long considered invariant, but is now known to vary on time scales from days to decades and probably longer. The mean value of the so-called solar constant is about 1367 W/m^2.

Surprisingly, at the height of the solar cycle (the sunspot maximum) when dark sunspots are most numerous on the solar disk, a "positive correlation" exists and the sun shines with a greater intensity. "Extra" energy leaves the sun's surface at a sunspot maximum from *faculae* (Latin meaning torches), bright areas surrounding active sunspots. How and why the energy gets from the sunspots to the faculae remains a mystery.

Perhaps even more critical than the 0.1% solar-constant changes are the variations in "spectral irradiance." The short wavelengths in the ultraviolet (UV) and extreme ultraviolet (EUV) vary more than 10% throughout the solar cycle. Although the research remains poorly understood, these variations can significantly influence the thinnest and most sensitive layers of the Earth's atmosphere and so may have important implications for climate change.

Even less well known are the longer-term influences of solar activity upon the solar constant. The record of earlier solar activity can be deduced from cosmogenic isotopes (^{10}Be, ^{18}O, ^{14}C, etc.) which show that Earth's temperature record often seems to correlate directly with solar activity: when this activity is high, the Earth is warm. During the famous "Little Ice Age" during the seventeenth century, the climate was notably cooler not only in Europe, but throughout the world. This correlated with the "Maunder Minimum" on the sun, an interval of few sunspots and aurorae (geomagnetic storms). In the eleventh and twelfth centuries, a "Medieval Maximum" in solar activity corresponded to the "Medieval Optimum" in climate, with global warming so prevalent that the Greenland Viking colony flourished. As solar activity declined, so did the global temperature, forcing the Vikings to retreat southward. At the end of the 1700s and the early years of the 1800s (the "Modern" or "Dalton Minimum"), solar activity dipped, and this era also proved cold. The twentieth century has been marked by generally increasing levels of solar activity. Cycle no. 19, peaking in 1958, produced the highest levels of sunspot activity recorded since Galileo's telescopic observations of sunspots in 1610. The 1990 peak appears to have been the second largest. This global temperature increase approximately parallels solar activity. Recent releases of Earth's greenhouse gases such as carbon dioxide have also caused a warming, so it is not clear how much of the warming can be attributed to each mechanism.

From an astronomical point of view, the sun is a mundane star. In this we are fortunate, because if the sun's variations proved too violent, Earth could not have provided a safe haven for the evolution of life, which requires great stability for hundreds of millions of years. Nevertheless, the sun displays a wide range of exciting astrophysical phenomena in interesting, but modest, variations: a hot corona with a temperature of millions of degrees, solar flares, sunspots, and faculae. In addition, the sun contributes significantly to Earth's natural climate variability.

A sunspot is a dark region on the sun (Figure 2.2). Although any individual sunspot covers only a small fraction of the solar disk, very large sunspots can have diameters up to about 10 times that of the Earth. Sunspots are dark because they are cooler than their surroundings and thus radiate less energy: however, their ability to stem the enormous flow of convective energy carried to the sun's surface is quite remarkable.

This chapter reviews sunspot observations from ancient accounts, through their telescopic discovery in 1610, to the modern era, and describes some key individuals and their observations. A chronological approach allows us to gain an appreciation for the slow development of new ideas in solar physics, ideas that often stymied theories about any possible sun/climate connections. Following this historical account, we will describe modern observational theories.

Pretelescopic Observations of Sunspots

The Aristotelian/Christian world view that the sun is a perfect body would certainly make anyone in Europe reluctant to report a sunspot. Several possible references to sunspots exist before the spread of Christianity. We have already noted Theophrastus' reference. The Roman poet Virgil (70–19 B.C.) wrote, "And the rising sun will appear, covered with spots." Charlemagne's astronomers supposedly saw spots on the sun in A.D. 807. The Arabic astronomer Abu Alfadhl Giaafar followed a sunspot for 91 days in A.D. 829. In A.D. 1198 Averroës of Cordoba mentions a spot on the sun, which he attributes to Mercury. In what may be only a fable, Joseph Acosta in his *Historia Natural des las Indias* published in 1590 in Seville supposedly states that the Inca Huyuna-Capac observed spots on the sun between 1495 and 1525. Modern solar studies suggest few sunspots existed during these years, casting some doubt upon Acosta's assertion. In 1607 Johannes Kepler saw a black speck on the sun, but, like Averroës, he attributed it to Mercury passing across the solar disk. The meagerness of the European naked-eye sunspot record may arise from two causes: (1) much of the ancient Greek and Roman scientific material was de-

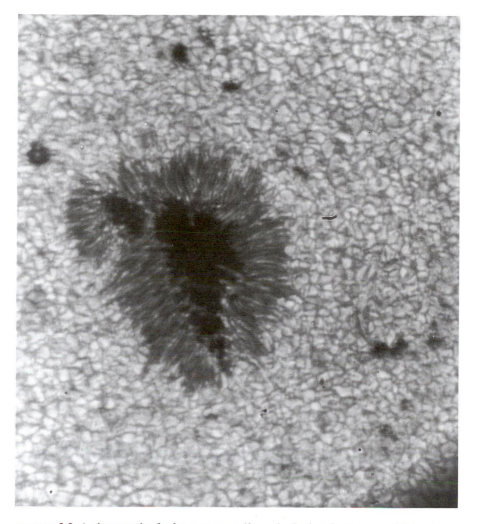

FIGURE 2.2 A photograph of a large sunspot (from the Project Stratoscope of Princeton University, with the permission of Martin Schwarzschild). A large sunspot can cover a billion square miles, or more than 700 times the surface area of the Earth. A sunspot's dark central portion is called the umbra. The lighter region surrounding the umbra is the penumbra. The sunspot is embedded in the photosphere. Convective cells (or granules), collectively known as granulation, surround the sunspot. Each granule is about the size of Texas and lasts about 10 to 20 minutes. The frontispiece to this volume shows a number of sunspots on the solar disk.

stroyed, and (2) the prevailing Christian world view tended to suppress naked-eye sunspot reports.

Naked-eye sunspot observations are more numerous in the Chinese chronicles, which date from around 800 B.C. During the last hundred years or so, many individuals have combed these records and discovered results so detailed that many aspects of solar activity can be traced back thousands of years. A more thorough discussion of these important discoveries is found in chapter 10.

The Discovery Controversy

It was a warm spring day in Padua in the year 1610. The telescope had been invented only 3 years earlier in Holland. Yet already replicas of this new marvel were spreading throughout Europe. One spring day, Galileo Galilei had turned his telescope toward the sun. (To avoid eye damage, we caution readers never to observe the sun directly through a telescope. Typically, astronomers project the solar image onto a surface from which it can be viewed.) A crowd of prelates, including Father Fulgenzio Micanzio and other men of letters, gathered to view the results. According to Micanzio, the sun's image was projected onto a white screen. At this time, most people believed the sun to be a perfect sphere. To the surprise of many, roughly a half-dozen dark blotches blemished the sun. What were these dark spots? Some thought there were defects in the telescope. Nevertheless, when Galileo rotated the telescope, the sun's image remained unaltered, proving the telescope was not the culprit. Others wondered if the spots were swarms of planets or objects passing in front of the sun. The more radical observers thought the spots were on the surface of the sun itself. By 1611 Galileo knew the answer. He had first observed the sun with a telescope in 1610 while still a lecturer in mathematics at the University of Padua. Yet because he was then embroiled in many controversies, Galileo wrote nothing on this subject in 1610 and 1611, but postponed his announcement, although he had indeed discovered sunspots.

Meanwhile, in Europe, others were also observing the sun. In December 1610, Thomas Harriot of Petworth, England, viewed the sun with his new telescope, first waiting until the sun was near the horizon and the air misty. Quick glances through the telescope enabled Harriot to examine the sun's disk. Harriot made the first known drawings of sunspots. For 199 days during 1611 and 1612 Harriot continued to view and draw the sun. As these drawings were made for his own benefit, his findings, like Galileo's before him, failed to attract world attention. In fact, his drawings remained unexamined until 1784.

In Germany, as in Italy and England, more telescopes were being turned toward the sun. In 1611 Johann Fabricus, the son of astronomer David Fabricus and a student at the University of Wittenberg, returned to his father's home in Osteel carrying several telescopes. That summer young Fabricus used his telescopes to examine the sun. Like Galileo and Harriot before him, he observed spots, and then he compiled his observations in a 22-page pamphlet entitled "De Maculis in Sole Observatis," published at Wittenberg (Figure 2.3). This pamphlet, the first publication on sunspots, was distributed at the Autumn Fair in Wittenberg in September 1611 and is listed in the Fair's Book Catalogue, which was widely distributed to the learned men of the day.

Fabricus's discovery provides an excellent account of the excitement generated by sunspots. The following translation from the German (by H. L. Crosby

JOH. FABRICII PHRYSII

De

MACULIS IN

SOLE OBSERVA-

TIS, ET APPARENTE

earum cum Sole conver-
sione,

NARRATIO

cui

Adjecta est de modo eductionis specie-
rum visibilium dubitatio.

VVITEBERGAE,

Typis Laurentij Seuberlichij, Impensis Iohan. Bor-
neri Senioris & Eliæ Rehefeldij. Bibliop. Lips.

....

ANNO M. DC. XI.

FIGURE 2.3 A copy of the cover of
Johann Fabricus's pamphlet covering
the first published account of sunspots.
(From Mitchell, 1915.)

of the University of Pennsylvania for Walter M. Mitchell) appeared in *Popular Astronomy* in 1915:

> While observing these things [i.e., the sun] carefully, a blackish spot suddenly presented itself, on one side indeed rather thin and faint, of no little size compared to the disk of the sun. I had at first no little doubt in the reliability of the observation, because a break in the clouds disclosed the rising sun to me, so that I thought that the clouds flying past gave the false impression of a spot on the sun. The observation was repeated perhaps ten times with Batavian telescopes of different sizes, until at last I was satisfied that the spot was not caused by the interposition of the clouds. However, not willing to believe in the manifest testimony of my own eyes, on account of the strange and unusual appearance of the sun, I immediately called my father, at whose house I was then staying, having returned from Batavia, in order that he might be present also to observe this. . . . Thus the first day passed, and we left the sun, but not without great longing for its return on the morrow, so that our natural curiosity scarcely bore even the intervention of the night. Nevertheless we restrained our eagerness by anxious thoughts. For it was not yet certain whether that spot which we had seen would wait for the next observation, which made us the more impatient the more uncertain we were in so great a matter. However, after having discussed the matter this way and that, each of us viewed the outcome according to his nature and desires. I, at all events preferred to doubt, rather than forthwith to form an opinion on the dubious testimony of a matter of uncertainty, which would have to be abandoned not without shame if the matter should turn out differently. Nevertheless I proposed myself two alternatives, one of which must be withdrawn from consid-

eration. For the spot either was on the sun, or was exterior to the sun. If on the sun there was no doubt but that it would be seen by us again, but if exterior to the sun it was impossible that it should be detected on the disk of the sun on successive days. For through its own motion, the sun would have moved away from this little cloud or body suspended between us and the sun. That night passed in doubting rather than in sleep; when we were aroused by the return of the sun which with its serene countenance rendered a welcome decision for us in that doubtful affair. Running, hardly bearing the delay of my curiosity to see the sun, I observed it. At the first glance of my eye the spot immediately appeared again, affecting me with no small pleasure. Since, although my doubt of the night before had prepared an alternative solution, by either of which we would learn the truth of the matter, still, by some intuition, I had secretly chosen this one. And thus it passed, we spent this day with frequent glances at the sun, scarcely satisfying our desires for observing, although our eyes with difficulty endured our persistence, which they protested against by threatening some great danger.

Although it was the first publication on sunspots, Fabricus's pamphlet received little widespread recognition, no doubt due to several factors. Apparently few copies of the pamphlet were published, so within a very short time it became a rare document. Johann Fabricus himself was not well known, so people ignored the work. But most important was the appearance of another writer, calling himself "Apelles," whose controversial claims pushed Fabricus's work into the background.

The Theory Controversy—Three Early Theories

As mentioned earlier, most people from this era considered the sun a perfect sphere. The teachings of Aristotle, adopted by the Catholic Church, maintained that a perfect sphere could not have blemishes. Basically, Aristotle believed that celestial objects were incorruptible, so sunspots could not be a solar phenomenon. Apelles, who was later revealed as a Jesuit priest named Christopher Scheiner, decided to defend the orthodox Aristotelian viewpoint. When Scheiner told his superior he was observing sunspots, his superior replied: "You are mistaken, my son. I have studied Aristotle and he nowhere mentions spots. Try changing your spectacles." In this intellectual atmosphere, Apelles began his discourse on sunspots with a public letter to Welser at Augsburg, who was a member of the nobility. In the first of three letters, Apelles argued that spots were not defects in observers' eyes because numerous people using eight different telescopes had noted the same number of spots in the same locations on the solar disk. Nor did revolving the telescope on its axis alter the results. Apelles then argued that the spots were not located in Earth's atmosphere, but rather were real bodies in or near the sun. Yet if they were in the sun, this would indicate that the sun rotates, contradicting the Aristotelian viewpoint. Apelles then logically concluded that the spots were bodies revolving around the sun. In the second letter, he argued that as Venus revolved around the sun, so did the spots. In the third and final letter, dated December 26, 1611, Apelles argued that because spots require 15 days to transit the solar disk, they should reappear after an equal interval. Failure to reappear is evidence that the spots are not part of the sun. He also suggested that the spots are near the sun and are probably swarms of small planets orbiting inside the orbit of Mercury. This became known as the "planetoid theory."

In these letters, Apelles also advanced this theory, which became popular between 1611 and 1635. Others argued that the spots were analogous to volcanoes on the Earth. Galileo was a proponent of the theory that the spots were similar to terrestrial clouds. In due course, Apelles revealed that he was Christopher Scheiner. His letters upset Galileo in at least two ways: Apelles was claiming (1) that he had discovered the sunspots and (2) that sunspots were not part of the sun, in contradiction to Galileo's own conclusions. Though Scheiner, Harriot, and Fabricus each independently discovered sunspots, historians have generally given Galileo credit for their initial discovery. It is reasonable to suppose that others also independently discovered sunspots in the years 1610 and 1611 but never recorded their findings.

Galileo rebutted Scheiner several times. As Galileo's viewpoints on sunspots are so correct and modern, it is worthwhile quoting him at length. In reply to Apelles' claims, Galileo stated:

> The dark spots, which are seen with telescopes on the disk of the Sun, are not far distant from it, but are contiguous in it, or are separated by such a small interval that it is imperceptible. Moreover they are not stars nor other solid bodies of long duration, for continually some are being produced and others are being dissolved; some being of short duration as 1, 2, or 3 days, and, others of longer duration as 10 or 15 days, and I believe others of 30 or 40 days or more. They are mostly of an irregular figure and they change shape continuously, some with rapid and large changes, others more slowly and with less variation; moreover, they increase or decrease in density, some appearing to condense and at other times to become rarified and diffuse; and besides changing into various shapes, one may be seen to divide into three or four and often many are united into one, which happens more often near the circumference of the solar disk than near the middle. Besides these irregular and individual motions of uniting and separating, condensing and changing figure, they have a maximum, common and universal movement, with which uniformity and in parallel lines they move over the body of the Sun; from the peculiarities of this movement it becomes known, first, that the Sun is absolutely spherical, second, that the Sun revolves on itself about its center bearing the spots with it in parallel circles and completing its revolution in about a lunar month, with a revolution similar to that of the orbs of the planets, namely from east to west. Moreover, it is to be noted that the majority of the spots seem to occur always in the same region or zone of the solar body, comprised between two circles corresponding to those which include the declinations of the planets, and beyond these limits as yet not a single spot has been observed, but all between these confines; so that neither toward the north nor toward the south do they appear to depart from the great circle of the Sun's rotation more than about 28° or 29°. The fact that we see them all moving as a whole with a common and universal movement is a sure argument that this movement can

have only one cause, and not that each one of them is going around the body of the Sun like a small planet at different distances and in different circles. Hence, we must necessarily conclude, either that they are all in a single sphere and like the stars are carried around the Sun, or that they are in the body of the Sun itself which revolves in its place and carries them with it. Of the suppositions, the second appears to me to be true, the other false.

All Galileo's conclusions about sunspots remain true today. The sun rotates in about 27 days, and the sunspots are carried along in this rotation. Sunspots occur in two zones lying both north and south of the solar equator and are transitory phenomena, with an average sunspot lasting about 6 days. The longest-lived sunspot ever observed occurred in 1919 and lasted 134 days. Most important of all, sunspots are indeed solar phenomena and not planetoids or asteroids.

For several years Scheiner resisted Galileo's conclusions and was supported by such individuals as Jean Tarde in France and Karl Malapert in Holland. From 1611 to 1633, Scheiner claims he observed the sun nearly every day. In contrast, after about 1612 Galileo seldom studied sunspots. To his credit, in time Scheiner altered his views and eventually agreed with Galileo. In 1630 Scheiner published his conclusions about sunspots in a 780-page opus entitled *Rosa Ursina* (Figure 2.4). This book, the first on sunspots, is dedicated to Paulus Jordanus II, Duke of Bracciano, of the house of Orsini. The title *Rosa Ursina* is meant to declare the sun "The Rose of Orsini." By most accounts the book is a poor one. In the late 1700s Delambre criticizes it by saying "few books are so diffuse and so void of facts," and then states there is not enough material for 50 pages. *Rosa Ursina* upset Galileo because Scheiner devotes 50 pages to attacking Galileo while also claiming undue credit for important discoveries. The book states that Scheiner spent 20 years studying the sun, making as many as 20 observations per day. Unfortunately, only a small fraction of these observations actually made their way into the book, and most of Scheiner's observations now appear lost.

The publication of *Rosa Ursina* ended a 20-year controversy about the nature of sunspots. Nonetheless, the book's publication may have had an unintended consequence, because the following decade produced fewer sunspot observations. Perhaps many of Scheiner's contemporaries viewed this book as the final word on the subject and turned their interests to other subjects. During the 1630s, entire years pass with no surviving sunspot records.

Early Observations to 1650

Galileo, Scheiner, Harriot, and Fabricus were not the only sunspot observers between 1610 and 1650. At least 30 more observers left written records of their observations. There are probably an equal number who studied the sun, but whose results were destroyed or misplaced during the intervening centuries. A few of these observers deserve more attention.

One of the earliest observers was Simon Mair, who wrote under the Latin name Marius. In 1619 Marius published an 18-page pamphlet devoted mostly

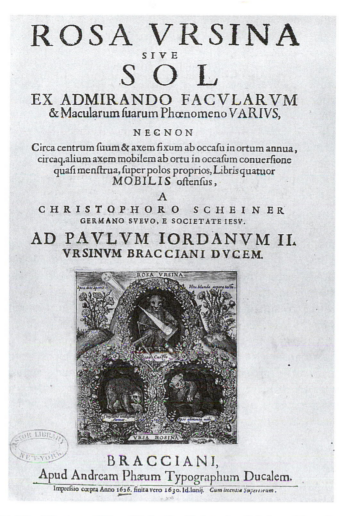

FIGURE 2.4 The title page of *Rosa Ursina* by Christopher Scheiner. (From Rare Books and Manuscripts Division, The New York Public Library, Astor, Lenox and Tildon Foundations, with permission.)

to the great comet of 1618. In his pamphlet, Marius noted that the number of sunspots had decreased markedly between the year of their initial discovery and 1618. Several commentators, such as Riccioli and Zahn, later stated that during 1618 entire months passed without any sunspots. Marius was the first person to note a change in the number of sunspots, but more than two centuries would pass before others achieved a real understanding of these variations. If Marius or other early observers had studied the variations in the number of sunspots over time, perhaps the 11-year activity cycle would have been discovered in the early 1600s. As noted earlier, Scheiner observed the sun nearly every day for more than 20 years. The data supporting the 11-year cycle existed, but there was no interest in cyclical observations.

Other observers were active for brief intervals in the early 1600s, including Jean Tarde of France who studied sunspots from 1615 to 1619 in the hope that he could prove sunspots were planets. Charles Malapert in Belgium observed sunspots from 1618 to 1626 and agreed with Tarde's conclusions. Between 1626 and 1629, Daniel Mogling in Darmstadt, Germany, tried to measure the solar rotation rate. Other observers during this time included Horrox and Crabtree in England; Castelli and Riccioli in Italy; Vander Miller in Belgium; Jungius, Saxonius, Smogulecz, Cysatus, Schickard, Hortensius, Quietanus, and Rheita in Germany; Hevelius in Danzig; and Octoul, Petitus, and Gassendi in France.

Three of these observers play an important role in our story. The first is Pierre Gassendi (1592–1655). Gassendi may be considered a philosopher and a scholar whose wide-ranging interests included astronomy. Gassendi's astronomical career is said to have begun in 1631 when, at the age of 39, he observed Mercury's transit across the face of the sun. During the next 15 years Gassendi observed the sun on an irregular basis, and we have records of 88 days when he observed sunspots. It is not his scattered observations, but Gassendi's influence on others that makes him important to us. In the 1630s when Johannes Hevelius was trying to decide whether to pursue his astronomical interests or try a different career, Gassendi helped persuade him to pursue astronomy. In 1645, Jean Picard became Gassendi's assistant. Hevelius and Picard proved to be two of the most active solar observers during the years 1650 to 1685. As their observations are crucial to our modern-day understanding of the sun, we now examine their individual stories.

Hevelius and Picard

Johannes Hevelius was born in 1611, one of 10 children, to a Danzig (now Gdansk, Poland) brewer. Peter Kruger taught young Hevelius both mathematics and astronomy. In 1630 Hevelius studied law at the University of Leiden, and in 1631 he visited London. From 1632 to 1634 he was in Paris where he made the acquaintance of Gassendi and the astronomer, Boulliau. At this time, Gassendi urged Hevelius to pursue astronomy rather than law. Nevertheless, in 1634 Hevelius returned to Danzig where he married and worked for 2 years in his father's brewery while pursuing legal studies. After observing a transit of Mercury on June 1, 1639, Hevelius avidly pursued astronomy. The 1639 death of Kruger, who for many years urged Hevelius to become an astronomer, appears to be the catalyst that made Hevelius enter the field.

Hevelius actively observed from 1639 to 1685 and died in 1687 at the age of 76. His main interest was the moon's geography, of which he produced detailed crater and mountain maps. Like most astronomers of this era, Hevelius was also interested in the location of stars and the distance and size of the planets, the moon, and the sun. Today we term these studies positional astronomy. An active writer, Hevelius's surviving letters number 12,000 pages. Of his roughly 10 books, *Selenographia, Cometographia,* and *Machinae Coelistis* may be considered major works that covered the moon, comets, and astronomical instruments and observations, respectively. In all three books one can find solar observations. For example, *Selenographia* contains many drawings of sunspots shown traveling across the sun.

Hevelius's sunspot observations are listed in a 30-page section of the 1,200-page second volume of *Machinae Coelistis.* Fewer than 100 copies of this very rare 1679 book exist. The solar observations, which occupy a very brief section, might easily be overlooked. In fact, most scientists who have studied solar activity and tried to reconstruct the sun's behavior have forgotten them.

The solar observations range from late 1652 through 1679. Hevelius's main interest in observing the sun was not sunspots per se, but rather finding the sun's height above the horizon. However, he did look for sunspots and commented on them when they were present. His 1652 comments on sunspots refer to only 2 days; in 1653 they are mentioned on 11 of the 92 observation days; in 1654 it is 4 of 71 days. For the period 1655 to 1659, Hevelius observed sunspots only 4 days in 1657. In 1660, he mentions sunspots on 30 of 96 days. Over 9 years, spots are mentioned for only 51 days. Such a low level of activity is completely different from today's solar behavior. The last full year with no sunspot activity is 1810. Since 1750, no two consecutive years have passed without some sunspots. This century averaged three to four sunspot groups per day, while ranging from zero to 25 groups on any particular day.

In 1661 Robert Boyle reported a sunspot group from May 7 to May 19. Jean Picard saw the same group on the same dates. Beyond that, no other reports are evident except those by Hevelius. Hevelius saw a spot group from February 22 to 26. The same group returned and was seen from March 12 to 22. In April, the spot (if present) produced no comment. Then in May Hevelius probably saw the same spot as Boyle and Picard, but only from May 12 to 19. Hevelius noted the same group again from June 10 to 12. The group returned in early July and yet again in late July and early August. From its appearance on May 9th to its last reported observation on August 7th, the observations lasted 91 days. Modern observations indicate only one sunspot group in about 250 lasts for four solar rotations. If the group seen in February is the same one that disappeared in August, this group would have lasted seven solar rotations, or 166 days. Of the 20,000 or more sunspot groups in the last century, none equals this for longevity. This remarkable fact deserves further comment. Not only were sunspots few, but those that did appear were durable. Most sunspots that appeared between 1660 and 1700 crossed the sun's entire disk, and about 10% lasted three or four revolutions. Today, fewer than 1% of the sunspots last that long. Thus, some solar changes may span hundreds of years. Few scientific measuring programs can cope with changes of this duration.

At the time, these observations were not considered particularly special, and Hevelius continued observing the sun until 1679. After seeing spots on 3 days in 1661, he reported no more spots until 1671. Now let us return to Jean Picard, who, after 1666, proved a much more active observer than Hevelius.

Born in 1620, Jean Picard became a Jesuit priest. On August 20–21, 1645, he assisted Pierre Gassendi in observing a solar eclipse, and he remained with Gassendi for 10 years. When Gassendi retired in 1648 and returned to Digne in the south of France, Picard went with him and later returned to Paris with Gassendi in 1653. It appears that, upon their return to Paris, Picard began actively observing the sun in an effort to calculate the solar diameter. Little is known of Picard's early solar observations except that, according to Keill who saw Picard's notebooks in 1745, from 1653 to 1665 Picard saw only one or two sunspots. If Picard's later activity is indicative of his earlier activity, then he was observing the sun about 100 days per year. From 1666 until his death in 1682, Picard's surviving records suggest that he observed the sun on every clear day. On August 11th, 1671, Picard stated he saw a sunspot—the first one he had seen in 10 years.

The Famous Sunspot of 1671

The sunspot of August 1671 caused quite a stir and led to several publications. Of his discovery, Picard said he "was so much the better pleased at discovering it since it was ten whole years since he had last seen one, no matter how great the care he had taken from time to time to watch for them." G. D. Cassini, who was then in charge of the Paris Observatory, commented: "It is now about twenty years since that Astronomers have not seen any considerable spots in the Sun, though before that time, since the invention of the telescopes, they have from time to time observed them." This comment indicates that Cassini was evidently unaware of Hevelius's observations. The editor of the *Philosophical Transactions of the Royal Society* footnoted Cassini's claim to suggest that it was more like 10 than 20 years.

Hevelius also observed this sunspot that, according to his records, was the first he had seen since 1661. However, he did not rush to join his colleagues in publishing his findings. Martin Fogel in Hamburg reported he had seen no sunspots since October 1661. Fogel, who was primarily a botanist, traveled quite a bit in the early 1660s, so it is difficult to assess the reliability of his statement. Both Siverus of Hamburg and Stetini in Leipzig also reported observations about this sunspot. About Stetini we know nothing except that he saw this sunspot. Although he probably made many solar observations, we hear about Stetini only because he observed this particular sunspot. Six known observers of this sunspot suggest the sun was intensely scrutinized during these years.

The intense excitement surrounding the observation of a sunspot is only one indication that the sun was actively observed and yet very free from sunspots. Here are several more reasons that support this viewpoint:

- On average, there are five known observers per year from 1653 to 1699.
- These five known observers, averaging 176 observation days per year, specifically detail the presence or absence of sunspots.
- Many known observers make general statements that no sunspots

were seen between two specified dates. Unfortunately, we do not have their specific days of observation, only the assurance that they were "diligent."

- An observed sunspot is often seen just as it rotates around the east limb of the sun. This tells us that a spot does not enter the sun much before it is observed, suggesting that people are observing the sun a large fraction of the time.
- The discovery of a new sunspot is often reported by a new observer whom we do not hear from again. Therefore many observations were being made of which we are no longer aware.
- Finally, sunspot drawings during this time can show remarkable detail, which tells us that telescopes were quite adequate (see Figure 2.5).

Other Observers

After Picard's death in 1682, Phillipe La Hire at the Paris Observatory continued Picard's observations until his own death in 1718. An even more active observer than Picard, La Hire made 200 observations in a typical year, so few sunspots went unobserved. In the years following 1671, sunspots were seen on a few days in 1672, 1674, 1676, 1677, 1678, 1680, 1681, 1684, 1686, 1688, 1689, and 1695. The most active years were 1676 and 1684, with sunspots visible on about 59 and 47 days, respectively. According to Maunder, from 1676–1677 a sunspot was observed through four solar rotations, and in 1684 Cassini and Kirch also followed a spot through four rotations. While John Flamsteed stated that no sunspots were seen between 1676 and 1684, our studies show they appeared on about 40 days between these years, demonstrating their rarity but not their nonexistence.

FIGURE 2.5 Sunspot drawings by Picard from his notebooks showing the dark inner umbra and the surrounding, less-dark penumbra in considerable detail. These drawings tell us that the telescopes used in the 1600s were of high quality.

In Cambridge, England, John Flamsteed was the Astronomer Royal from 1676 until his death in 1725. From 1676 to 1699 he frequently observed the sun, making about 60 or 70 solar diameter measurements each year. Today Flamsteed's correspondence and papers are in the Cambridge University Library. Two comments from his unpublished letters are of interest here. In one letter he writes, "As for spots in the Sun there have been none since the year 1684. You may acquaint Mr. Ayres of it and that which is published in foreign prints is a romance. The Sun having been as clear of late years as ever, and I have seldom omitted observing him at noon when it was clear." Although some popular literature in that era discussed sunspots, exactly what was said remains unknown to us. In a follow-up letter, Flamsteed says: "I told you in my last [letter] no spots have been seen in the Sun since 1684. All the stories you have heard of them are a silly romance spread as such as call themselves witty men to abuse the credulous and [are] not to be heeded." Flamsteed is quite correct here. The only sunspots observed between 1690 and 1699 occurred on four days in late May 1695 by La Hire in Paris and Maraldi in Bologna. La Hire's comment on this spot is, "It is a long time since anything so great as these have appeared."

A Table of Sunspots Seen from 1672 to 1699

For the most part, the near absence of sunspots in the last half of the seventeenth century was later forgotten or dismissed. In 1726 Chr. A. Hausen noted that no spots were observed from 1660 to 1671 and from 1676 to 1684. Others also made occasional, rare comments to this effect. For example, in 1796 and 1801 Sir William Herschel thought the telescopes during this earlier period were inadequate to see sunspots. In 1942 W. A. Luby thought perhaps observers seldom observed the sun. In 1889 Gustav Spoerer at the Royal Leopold-Caroline Academy published two articles showing there was a real dearth of sunspots in the late 1600s. E. Walter Maunder, working for the Royal Greenwich Observatory in England, considered these results so important he translated them from the German and presented them in the Annual Report for the Royal Astronomical Society for 1890. He elaborated on his earlier accounts in the magazine *Knowledge* (1894). Even so, the results seem to have attracted little attention. In 1922 Maunder again tried to draw attention to Spoerer's results with an article in the *Journal of the British Astronomical Society*. In addition, Annie Maunder, who was E. Walter Maunder's wife, discussed the absence of sunspots in her book *The Heavens and Their Story*. Despite these efforts, the subject received little attention. S. B. Nicholson wrote about it for the Astronomical Society of the Pacific (1933). In the *Journal of the British Astronomical Society* (1941), H. W. Newton and P. Leigh-Smith noted the sun behaved unusually in the late 1600s. Luby, on the other hand, criticized this idea in *Popular Astronomy* (1942). Scant notice was taken in the professional literature, however. All this changed following a historically important 1976 study by John A. Eddy of the High Altitude Observatory. Eddy not only generated interest in the paucity of sunspots from 1645 to 1715 but also gave it the catchy name the "Maunder Minimum."

We will return later to the Maunder Minimum and this anomalous solar behavior to see what sense, if any, we can make of it and its possible influence on Earth's climate. Our discussion of the Maunder Minimum closes with part of a table first generated by Spoerer and translated by Maunder listing all the known observations of sunspots from 1672 to 1699 and summarizing our discussion.

Spoerer's Table of Sunspots (translated by Maunder)

1672   Nov. 12–13, South Latitude 13°.

1674   Aug. 29–31

1676   June 26–July 1, S. Lat. 13°; Aug. 6–14, S. Lat. 6°; Oct. 30–Nov. 1; Nov. 19–30; and Dec. 16–18. Three returns of the same, S. Lat. 5°.

1677   Same spot observed in fourth rotation; another April 10–12.

1678   Feb. 25–March 4, S. Lat. 7°; May 24–30, S. Lat. 12°.

1680   Spots observed in May, June, and August.

1681   Spots observed in May and June.

1684   Kirch and Cassini observed a spot through four rotations, April–July, S. Lat. 10°.

1686   April 23–May 1, S. Lat. 15°; Sept. 22–26.

1687   Cassini could find no spots, though observing carefully.

1688   May 12, S. Lat. 13°.

1689   July 19–22; Oct. 27–29; these spots are reported as ephemeral.

1689   March to May 1695. De La Hire reports that he found no spots.

1695   May 27, De La Hire says: "It is a long time since anything so great as these have appeared." No spots until November 1700.

1699   Last year wholly without spots.

In contrast, with fewer than 50 sunspots listed from 1672 to 1700, during the last century (1895–1995), a typical 30-year interval reveals 40 to 50 *thousand* sunspots.

In 1711 William Derham provided a striking one-sentence summary of the Maunder Minimum: "There are doubtless great intervals sometimes when the Sun is free, as between the years 1660 and 1671 and 1676 and 1684, in which time, Spots could hardly escape the sight of so many Observers of the Sun, as were then perpetually peeping upon him with their Telescopes in England, France, Germany, Italy, and all the World over, whatever might be before from Scheiner's time."

Return of the Sunspots in 1715–1716

From 1700 to 1710 sunspots were observed every year. For each of these 11 years there were 10 or more days during which spots were seen. In 1705 and again in 1707 sunspots were seen on more than 100 days. Yet in almost every instance, the solar disk contained only one group of sunspots at any time. By modern standards the sun remained subdued. From 1672 to 1705 all the sunspots were in the southern hemisphere. Then in April 1705 a spot appeared in the sun's northern hemisphere.

These solar changes generated widespread interest. In the first decade of the 1700s, at least 25 people started to make and subsequently record or publish their sunspot observations. No previous decade had created such curiosity. Many other individuals undoubtedly observed sunspots, but failed to record their sightings. Three observers deserve special mention. One is Reverend William Derham of Upminster, England, who observed from 1703 to 1715. Derham was an active observer whose primary interest was sunspots rather than solar diameter or solar rotation measurements. Many of Derham's results were published in the *Philosophical Transactions of the Royal Society,* but to this day some of his unpublished observations remain in the Cambridge University Library. On September 10, 1714, Derham commented that "no spots have appeared on the Sun since October 18, 1710." Other observers, such as La Hire and Wurzelbaur, confirmed this statement.

Another noteworthy observer is François de Plantade of Montpellier, France, who was active from 1704 to 1726. His observations were never published, and the present location of his observations is unknown. Plantade's results were recorded in the 1860s scientific literature, so we have some idea of what he saw. Plantade was also the most active observer in 1726.

The final outstanding observer of this era is Stephen Gray of Canterbury, England. Like Derham, Gray made his observations at Cambridge. On December 27, 1705, Gray reported seeing a "flash of lightning" near a sunspot. Today we call this phenomenon a white-light flare, an explosive release of energy rarely seen in the visible light spectrum. Most common flares affect only the thin upper portions of the sun's atmosphere, and then just the chromospheric emission lines. The next recorded white-light flare occurred in 1859 and was reported in the scientific literature by Richard Carrington. Gray's discovery remained in his notes, and at the time its significance went unremarked. Yet it is one more piece of evidence that the sun was changing. We now know that coronal mass ejections may be somewhat more important than flares in affecting the Earth's environment.

After nearly four quiet years from 1710 to 1714, sunspots returned to the sun with a vengeance. The subsequent years produced not just one group of sunspots, but two, three, four, or even five simultaneous groups. Fontanelle in Paris commented that the appearance of two groups of sunspots at once was unprecedented. In 1716 more than 160 days had sunspots; in 1717 more than 280 days. Such levels of activity may have had an unintended effect. In 1718 the sun was observed less frequently than it was in 1717. We could find no record that anyone even examined the sun during the months of June and July 1718, the first time an entire month passed without known observers since 1674, and the first year with two consecutive missing months since 1652. Were astronomers becoming bored by something that was now commonplace? Perhaps. We will return to this subject shortly.

In April 1716, for the first time in many years, three simultaneous sunspot groups appeared on the solar disk. On March 17, Kirch in Berlin reported two sunspot groups, the first time that year that more than one group had appeared. That same night the brilliant aurora borealis, or northern lights, was visible throughout Europe. Then people did not connect aurorae with the sun, but today we know aurorae are caused by magnetic activity on the sun.

The March 17, 1716, aurora was visible in London, throughout Germany and Prussia, and as far south as Italy. The last aurora to be seen so far south occurred in 1621, so ordinary people were greatly alarmed by the wonder in the skies. The almanacs called them the "Great Amazing Light of the North." In Scotland the aurora appeared on the night before Lord Derwentwater's execution, and long afterward they were known as "Lord Derwentwater's Lights." In Ceylon their name was "Buddha Lights." The Chinese created many engravings of the aurora. In Europe the aurora's appearance stimulated a few scientific articles. *Acta Eruditorum* in 1716 discusses the aurora in some detail, and Figure 2.6 reproduces an engraving showing their appearance. From these drawings one senses how alarming and strange aurorae were thought to be. In A.D. 555 Matthew of Westminster described aurorae as "lances in the air." Figure 2.7 shows a modern photograph of the aurora. In the *Philosophical Transactions* of 1716, the famous astronomer Edmund Halley described aurorae and developed a magnetic theory to account for them. The French government commissioned him to study and report on the aurorae, leading to his book *Traite de l'Aurora Borealis*.

Although the aurorae engendered panic among people with strong religious beliefs who considered them supernatural, as aurorae became more common they were considered more benign. In later years, for example, the Shetland Islanders called aurorae the "Merry Dancers." For us, however, the aurorae are important because they reveal something about the mean level of solar activity. Although several scientists made comments on a possible connection between aurorae and solar phenomena, it was not until 1850 that this connection was truly appreciated.

Returning to sunspot observations, from 1700 to 1718 the most active solar observer at the Paris Observatory continued to be Phillipe La Hire. When La Hire died on April 21, 1718, his son, Gabriel-Phillipe La Hire assumed his duties until he died on April 19, 1719. After these two deaths, for a time systematic solar observations by professional astronomers essentially ceased.

Observations from 1719 to 1761

After the younger La Hire's death in 1719, two medical doctors became the most active solar observers. One was J. L. Rost who observed for 2 years, and the other was Dr. J. L. Alischer of Jauer (today Jawor, Poland) who was active for many years. Except for being a prolific writer, largely for the journal *Sammlung von Natur und Medizin,* we know very little about Alischer. While most of his writings concern medical topics, in 1719 he began publishing portions of his sunspot diary called "Diarium Solarium Macularum." During 1720, 1721,

FIGURE 2.6 Drawings of the northern lights seen throughout Europe in 1716. (From *Acta Eruditorum.*)

and 1722 his observations provide the best contemporary record of solar activity. Yet his results were not published in 1723, and we know the sun's condition for only 9 days during that year, based on the scattered comments of five observers. What was happening? Alischer was obviously busy observing the sun, as he continued publishing portions of his diary in later years. We suspect that in 1723 the sun had very few sunspots and the editor decided publishing a string of null results might bore his readers. Although he was observing as late as 1727, when the journal *Sammlung* ceased publication in 1725, Alischer no longer had an obvious outlet for his work. Since his original diary may now be lost, we do not know the full story of what he saw.

FIGURE 2.7 A photograph of an aurora over Michigan. (From Richard A. Goldberg, with permission.)

With surviving records for only 9 days of solar observations in 1723, interest in the sun and sunspots was waning. Solar activity did increase to a maximum around 1726–1728, which caused Plantade to be very active in 1726. Yet even he ceased observations and retired in 1726, although he lived another 20 years. Following Plantade's retirement, the sun was apparently observed less than 100 days per year during each of the next 22 years. In 1734 Adelburner, a Nuremberg printer, reported that an anonymous observer had stated there were no sunspots seen in 1733. However, 1734 proved an extremely momentous year for solar astronomy, with no recorded solar observations by anyone. Even Rudolf Wolf, who searched the literature for 45 years, could find no recorded observations for this year. For the first time in more than 120 years, the sun held no interest for either scientists or amateurs. Why?

Comments by professional astronomers provide some clues. In 1739 Keill wrote that sunspots showed no constancy in their appearance or disappearance. Cassini reinforced this viewpoint in 1740 by saying, "It is evident from these reports that there is nothing regular about sunspot formation, their number, or their figure." In 1764 Long essentially repeated Cassini, stating, "Solar spots observe no regularity in their shape, magnitude, number, or in the time of their appearance or continuance." Since much of the excitement in astronomy involves the observation or discovery of regular, repeatable, or predictable phenomena, this discouraging attitude among the professionals seemed to imply that sunspots were not worth studying.

The situation became worse. In his sunspot research, published in 1868, Wolf could find only one observation in each year for 1738, 1740, 1741, 1746, and 1748. No observations were found for 1744, 1745, and 1747. For the 4 years from 1744 to 1747, there is only a single observation by Hallerstein, a Jesuit missionary in Peking, China.

Today a few more observations have been found, including about 70 days of observations made by Masuno in Venice between 1739 and 1742. Masuno was mainly interested in measuring the solar rotation rate. He appears to be the most active observer between 1734 and 1748. If additional observations still exist in some obscure archive, no one has located them.

Starting in 1748, solar observations increased slightly. Nevertheless, until 1800 observations remain fewer than desired. This era's major observers (with their starting and ending observation dates given) are J. C. Staudacher (1749–1799) of Nuremberg, L. Zucconi (1754–1760) of Venice, J. C. Schubert (1754–1758) of Danzig, C. Horrebow (1761–1776) of Copenhagen, P. Heinrich (1781–1820) of Munich, H. Flaugergues (1788–1830) of Viviers, and J. G. Fink (1788–1816) of Lauenburg.

We next focus on several individuals who made observations or discoveries important to our present understanding of the sun.

Observations of Christian Horrebow

Christian Horrebow was born in Copenhagen in 1718, the son of the astronomer Peder Horrebow. After receiving a master of science degree from the University of Copenhagen in 1738, Christian became his father's assistant at the Round Tower Observatory. Horrebow's main work consisted of compiling almanacs and measuring stellar positions. He was also a professor of astronomy and wrote textbooks in astronomy and mathematics. In 1761 he began systematically observing sunspots and continued to do so for the next 15 years. These observations are important today because in his own time Horrebow was about the only active sunspot observer. His daily observations, some 200 sunspots per year, exist for 1761 and 1764 to 1776. Earlier observations from 1762 and 1763 now appear to have been destroyed. In 1859 T. N. Theile tabulated Horrebow's monthly mean sunspot group numbers, and a few years later d'Arrest provided a different tabulation. Examining these two tabulations shows their counts of sunspot groups differ by about a factor of 2. D'Arrest's tabulation is probably

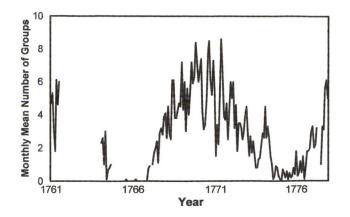

FIGURE 2.8 The monthly mean number of sunspot groups observed by Christian Horrebow and his colleagues from 1761 to 1777 based on the author's examination of the notebooks at the University of Aarhus. In 1873, Professor d'Arrest examined a portion of the notebooks and obtained very similar numbers. Thiele, in 1859, on the other hand, called individual sunspots "groups" and so obtained very high counts. Thomas Bugge and Erasmus Lievog made the observations after Horrebow's death in 1776. It is surprising that from this data the 11-year solar cycle was not discovered earlier.

more nearly correct, but let us now examine Thiele's summary plot of Horrebow's observations (Figure 2.8).

It is surprising that Christian Horrebow did not discover the 11-year solar cycle from his own observations. Some argue that Horrebow did indeed discover this cycle but never published the result. Speaking about sunspots in his notebooks, Horrebow says their more systematic observation might lead to "the discovery of a period, as in the motions of the other heavenly bodies." He elaborates that "then, and not until then, it will be time to inquire in what manner the bodies which are ruled and illuminated by the Sun are influenced by the sunspots." From these comments, we cannot say that Horrebow discovered a periodicity in sunspots, although he had the data and the opportunity. With Horrebow's death in 1776, the new director of the Round Tower Observatory ended the systematic observations of sunspots. That change in priorities resulted in a missed opportunity to make a major new discovery about the sun.

The Wilson Sunspot Depression

Alexander Wilson was born in Edinburgh, Scotland, in 1714 and died there in 1786. Wilson is famous for his 1774 discovery of the "Wilson Depression" in sunspots (Figure 2.9). Since their discovery, sunspots were known to consist of two components—a dark inner region called the umbra and a lighter surrounding region called the penumbra (see Figure 2.2). This terminology is based on the similar darkness contrast of shadows that are darkest at their cen-

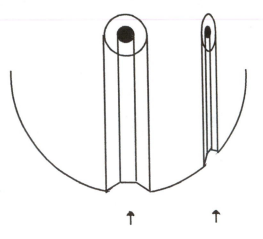

FIGURE 2.9 The Wilson sunspot depression. Sunspots may be viewed as depressions in the solar disk, as shown by a semicircular cross-section of the sun. The viewer is located at the bottom of the page and is looking in the direction of the arrows. At the top of the figure, two schematics of sunspots are shown as seen by the viewer. Because sunspots are depressions, the dark umbra of the sunspot appears offset away from the limb of the sun as the sunspot approaches the edge of the sun. Thus, penumbrae are compressed on the near side and expanded on the far side. The amount of sunspot concavity is exaggerated in this figure for illustrative purposes.

ter and less dark toward the edge. Wilson noted that as sunspots approached the limb of the sun, the penumbra on the side of the sunspot nearest the limb was broader than the penumbra nearest the center of the disk. Wilson explained this as a geometrical phenomenon and stated that the umbra could be viewed as depressed below the normal surface of the sun, or that spots are concave depressions. Thus as the spot moved toward the limb, it would be easier to view the penumbra on the far side of the spot than the penumbra nearer to the observer.

Wilson's observations were another step along the road to understanding sunspots. Many observers tried to disprove Wilson's findings, and, while none succeeded, these additional observers left behind numerous sunspot observations that might otherwise never have been made.

Sir William Herschel and the Revival in Sunspot Observations from 1800 to 1826

William Herschel was born in Hanover, Germany, in 1738, one of 10 children, five boys and five girls. His father, a musician, had a musical ensemble. At age 14, Herschel was playing the hautboy and learning the violin. In time he would learn the harpsichord and the organ. When Herschel was 19 and a member of the Prussian army, the Seven Years' War started. Rather than go to war, with the help of his mother and sisters he boarded a trading ship, and in 1757 he arrived in England, with only a single crown, and made his way to London to begin a musical career.

Eventually Herschel went to Bath, a town that was, at that time, the entertainment capital of England. For a time he played the organ there, but he wanted a vocalist to accompany his concerts. Finding no satisfactory vocalist in England, Herschel returned to Hanover and convinced his sister Caroline to be his singer. Although she knew no English, within a year Caroline was singing solo parts in William's concerts. On his own, William might have continued his musical career until his death, but after their concerts William and Caroline took long night walks during which they studied the stars and talked of Kepler and Copernicus. On their nights off, Herschel often studied Smith's Harmonics or Ferguson's Astronomy. To better observe the stars, Herschel wanted his own telescope. In a second-hand store they found a 2-inch Gregorian reflector. Then they started making a larger telescope, constructing the tube from pasteboard, layering sheets of paper from old books and letters, and occasionally using cloth strips. This telescope eventually allowed them to see Saturn's rings. One night William and Caroline left their telescope outdoors, and it rained. The next morning the telescope was bent in a circular arc and only the lenses and mirrors could be recovered. Rather than feeling defeated and giving up astronomy, William and Caroline immediately drew up plans for a better telescope. At first they offered to pay an instrument maker 50 pounds (a considerable sum in those days) to make a 6-foot-long telescope. He refused. Though no one accepted their offer, an amateur lens maker did sell them all his tools. Now they had a machine shop.

For the next 7 years, Herschel constructed telescopes and viewed the heavens from outside his quarters in Bath. One night, to get away from the shadows of the houses, he moved his telescope to the middle of the road where a gentleman named Sir William Watson chanced by. Watson asked to look through the telescope, and one look convinced him that Herschel's optics were superior to other telescopes of the day. Watson was to open the doors of the scientific community to Herschel. In March 1781 Herschel found the planet Uranus. His great discovery, the first new planet ever found, made Herschel famous at age 43. Herschel was nominated by his old friend Watson to become a member of the Royal Society, and he later received honorary doctorates from the universities at Edinburgh and Oxford. In 1782 George III appointed Herschel the "Astronomer to the King," and Herschel moved from Bath to London to live near Windsor Castle. The 100 pounds per year provided by the king allowed Herschel to pursue his astronomical interests in earnest. Caroline was given the title "Assistant to the King's Astronomer" along with 50 pounds per year.

Although primarily interested in the stars and the night sky, starting in 1800 Herschel turned his attention more toward the sun. His initial motivation was the possibility of finding a planet closer to the sun than Mercury, the so-called intramercurial planet. Soon, however, sunspots, faculae, and the solar granulation captivated his attention. Herschel believed sunspots were openings in the sun's luminous atmosphere. Through the solar clouds Herschel imagined he was seeing a cooler, darker surface. He also hypothesized that the sun should be considered a planet, perhaps even an inhabited sphere. To us this seems absurd, but in Herschel's day light and heat were not necessarily considered the same. An object like the sun could be bright yet cool because a bright light might not generate heat until the light was absorbed. More importantly, Herschel believed that, since sunspots were cool, a change in their area or number could lead to a change in the sun's light emission and hence could affect the weather. His thoughts were not original on this topic, but because he was so famous, his comments stimulated interest in both sunspots and in a sun/weather relationship which increased during the following years. During 1801 Herschel published two papers in the *Philosophical Transactions of the Royal Society* on the sun as a variable star, entitled "Observations tending to investigate the nature of the Sun, in order to find the causes or symptoms of its variable emission of light and heat" and "Additional observations tending to investigate the symptoms of its variable emission of light and heat of the Sun, with trials to set aside darkening glasses, by transmitting the solar rays through liquids and a few remarks to remove objections that might be made against some of the arguments contained in the former paper." Among other topics, Herschel announced the discovery of light outside the eye's own visible range, now known as infrared radiation. We will return to Herschel's thoughts concerning sunspots, wheat prices, and weather in a later chapter.

Herschel's observations stimulated a revival of interest in sunspots. His first paper dealing with the sun appeared in 1795, and every subsequent year from then onward the sun was observed for more than 100 days. Flaugergues, von Ende, Fritsch, Derfflinger, Stark, Tevel, Adams, Pastorff, Arago, and Schwarzenbrunner are to be counted among those who studied sunspots during the next quarter of a century.

Heinrich Schwabe and the Discovery of the Eleven-Year Cycle

Samuel Heinrich Schwabe was born in Dessau, Germany, in 1789 and died there in 1875. His father was a physician, and his mother's family ran an apothecary or drugstore. From 1806 to 1829 Schwabe either worked or trained to be an apothecary. At the age of 40, he sold his interest in the apothecary and devoted himself full-time to his scientific interests.

Like Herschel before him, Heinrich Schwabe studied the sun in hopes of discovering an intramercurial planet. Starting in 1825 Schwabe began searching for an intramercurial planet, but instead he soon began examining sunspots. From the beginning of 1826 until 1868 he meticulously recorded the appearance and disappearance of sunspot groups. We have already mentioned how Keill, Cassini, and Long considered studying sunspots pointless. In later years, the famous French astronomers Delambre and Lalande reiterated these same negative views. Despite this, during most years Schwabe viewed the sun on 300 or more days, although some years he recorded fewer than 200 days. Summaries of his results, usually the number of days without spots and with spots each year and month and the number of new sunspot groups, were published in *Astronomische Nachrichten*. After 17 years of careful observations, in 1843 Schwabe announced the discovery of a 10-year cycle in the number of sunspots. For the most part, his findings went unnoticed and unremarked. Rudolf Wolf in Bern, Switzerland, is likely to have noticed Schwabe's conclusions because by 1847 Wolf also was actively observing the sun and beginning to comb the writings of previous astronomers to find old sunspot observations. Another person who noticed Schwabe's conclusion was Alexander von Humboldt, a famous naturalist and renaissance man. Humboldt's best-selling book, *Cosmos,* published in 1851, described Schwabe's discovery and brought Schwabe widespread attention. In 1857 Schwabe received a gold medal from the Royal Astronomical Society, and in 1868 he was elected a member of the Royal Society. Instead of the intramercurial planet he first sought, Schwabe instead found the 11-year, or Schwabe sunspot, cycle (Figure 2.10). When receiving these awards, Schwabe said of his efforts, "I can compare myself to Saul, who went out to find his father's asses and found a throne."

Rudolf Wolf and the Wolf Sunspot Number

Rudolf Wolf was born near Zurich, Switzerland, in 1816, the son of a minister (see Figure 2.11). After receiving an education in astronomy in Zurich, Vienna, and Berlin, he moved to Bern, Switzerland, to teach mathematics and physics. He began teaching astronomy in 1844, and in 1847 he started observing and recording sunspots. Already well read in the historical literature and inspired by Schwabe's discovery of the sunspot cycle, Wolf wondered if there were sufficient historical sunspot observations to extend the cycle's shape and timing

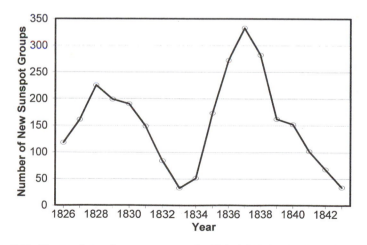

FIGURE 2.10 Observations of sunspot groups by Heinrich Schwabe through 1843 when he announced the discovery of the 11-year solar cycle. The number of newly appearing sunspot groups each year are plotted here.

FIGURE 2.11 A portrait of Rudolf Wolf in 1855. (From Wissenschaftshistorische Sammlungen der ETH-Bibliothek, Zurich, with permission.)

to earlier periods. As early as 1852, Wolf could affirm that the sunspot cycle was not 10 years but rather 11.111 years. Most of these results and supporting background material were published in Wolf's own solo-authored journal *Astronomische Mittheilungen.* These 13 volumes, dating between 1852 and 1893, contain a gold mine of data about sunspots and much information that is unavailable elsewhere. After 1855 Wolf lived in Zurich where he taught and conducted his research (Figure 2.11).

To his credit, Wolf's statement that the sunspot cycle averaged 11.111 years did not blind him to the fact that some cycles were only about 8 years long while others appeared to last 17 years. By the early 1850s Wolf had designed an index called the sunspot number and now known as the Wolf, or Zurich, Sunspot Number. This index is simply 10 times the number of sunspots plus the number of individual sunspots all multiplied by a constant that differs for each observer. The constant or personal equation puts all the observers on an equal footing, and it corrects for differences in telescopes and in the diligence of these same observers in recording small individual sunspots or sunspot groups.

Sunspots seldom appear alone but generally cluster together in groups. A common type of group consists of two nearby spots traveling together across the solar disk. Today this group would be called bipolar because each spot has a magnetic polarity. The two individual spots are designated as the preceding (p) and following spots (f), and they have opposite polarities (see Figure 2.12). They might be accompanied by smaller spots that appear and disappear during the life of a spot group. A group can range from a single isolated spot to a complex grouping of 30 or 40 individual spots. Intense solar activity sometimes makes it difficult to distinguish one very large group from two nearby groups. Some ambiguity exists with the definition of groups, so observers disagree about the actual number of groups, but a general agreement exists between most observers. The sunspot group is generally well defined, and even the earliest observers noted its existence.

By 1868 Wolf had reconstructed solar activity after 1700. His reconstruction was revised in 1873 and has since been updated by the Zurich Observatory. The numbers are often called the Zurich Sunspot Numbers. Another new reconstruction of solar activity has also been made using only sunspot groups, called Group Sunspot Numbers. The Group Sunspot Numbers are very similar to the Wolf Sunspot Numbers published in 1868 (see Figure 2.13). Motivated in part by criticisms of Elias Loomis, an American who in 1873 pointed out that for some years in the early 1800s Wolf had only 1 or 2 days of observations, Wolf acknowledged a problem. Wolf bypassed the problem by using auroral and magnetic needle observations to help fill in those years when telescopic observations were sparse. By these techniques Wolf increased his numbers before 1848 by 25% to 50%.

What could justify Wolf's auroral and magnetic needle substitution procedure? With Schwabe's announcement of his sunspot cycle, other scientists began seeking similar cycles in other phenomena. In 1850 Alfred Gautier, Edward Sabine, and Rudolf Wolf simultaneously and independently announced the existence of an 11-year cycle in the movements of magnetic needles. For years people had observed these needles and recorded their results. Some days the

Magnetic polarity for bi-polar sunspots in one solar cycle
in one particular hemisphere (north or south).

Magnetic polarity for bi-polar sunspots in the next solar cycle
in the same hemisphere.

FIGURE 2.12 A schematic diagram of bipolar sunspots showing how they change their magnetic polarities from one solar cycle to another. (Inspired by Stetson, 1947.)

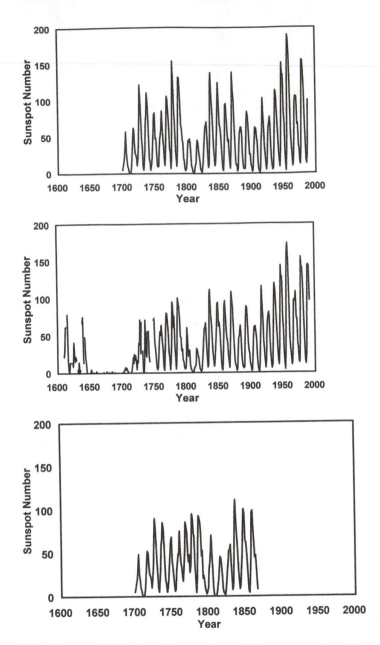

FIGURE 2.13 Three panels showing different solar activity reconstructions. The top panel shows the Wolf Sunspot Numbers commonly used in many climate and solar studies. The middle panels shows the Group Sunspot Numbers. They are nearly identical to the Wolf Sunspot Numbers after 1848. Before that time they are 25% to 50% lower. In the lower panel, the Wolf Sunspot Numbers published by Wolf in 1868 are shown. They are nearly identical to the Group Sunspot Numbers. Wolf's revision of these numbers was probably incorrect, so the middle panel probably represents the best reconstruction of solar activity. The two prolonged decreases in sunspot numbers are known as the Maunder Minimum and the Dalton Minimum, respectively.

needles showed more disturbance than others, and some years showed more disturbed days than others. At the time the cause was unknown, but today we know the sun has a magnetic field and a constant outflow called the solar wind. A disturbance on the sun propagates through the interplanetary medium, altering the Earth's magnetic field and hence magnetic needle observations. The more active the sun, the more disturbances and needle changes there are. Gautier, Sabine, and Wolf all noted the correlation between solar activity and the magnetic needle fluctuations. These observations could be used to help reconstruct solar activity because the needle observations went back to about 1784.

Wolf used these needle observations to adjust the amplitude of his sunspot reconstruction. He was almost forced to do so because for many years direct observations of sunspots were quite scarce. Figure 2.14 tabulates our own frequency of sunspot observations as a function of time. For many years between 1740 and 1780, the interest in sunspots was low, with few recorded observations. One can get past this problem by using magnetic needle and auroral observations, but doing this introduces other problems that Wolf apparently did not fully appreciate. The most serious problem is that auroral numbers tend to peak 2 or 3 years after sunspot numbers, so aurorae are not a perfect proxy for sunspot numbers. Figure 2.15 shows the long-term variations in auroral records. Loomis and Wolf both independently discovered the correspondence between the number of aurorae and the number of sunspots. Like so many similar discoveries, the earlier scientific literature contains hints that other individuals had noted this same relationship. Elias Loomis also discovered the auroral oval.

Wolf died on December 6, 1893, but his Zurich successors continued to update the records of sunspot numbers. Since 1981, people in Brussels have continued this work. So today, thanks to Wolf's original heroic diligence, we have almost 300 years of sunspot numbers from 1700 to the present. (Note,

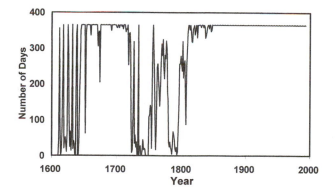

FIGURE 2.14 How frequently sunspot observations have been recorded for the sun (the number of days each year that have observations). Wolf's sunspot reconstruction used less data than shown here since he had few sunspot observations before 1750 and could only estimate yearly means.

however, that, due to lack of space, one of Wolf's successors destroyed much of Wolf's correspondence and perhaps other documents.)

Richard Carrington, Gustav Spoerer, and the Butterfly Diagram

With the discovery of the sunspot cycle, one might think almost everything important about sunspots had been found. Actually, many more revelations remained to be made.

Richard Carrington, the son of a beer brewer, originally intended to become a minister. At Trinity College in Cambridge, England, Humboldt's report of Schwabe's discovery of the sunspot cycle aroused Carrington's interest. Being nearly independent financially, in 1853 he began observing sunspots from his own observatory, and over an 8-year span he made three important solar discoveries.

The first discovery is the sun's differential rotation. Using sunspots as tracers in the sun's atmosphere, Carrington found that the equatorial regions rotated faster than the polar regions. Figures 2.16 and 2.17 show this phenomenon, known as the differential rotation.

Differential rotation is surprising for several reasons. The sun is a fluid with convection cells. Consider a fluid element at a certain latitude. Let us imagine the turbulent motion moves this element toward the equator. A Coriolis force develops, which slows the element down relative to its neighbors. Thus, as one examines regions closer to the equator, less rapid rotation would be expected to occur to conserve angular momentum. Yet the sun's equatorial regions rotate faster, rather than slower, than the higher latitude regions!

The second discovery about sunspots concerns their creation in various solar latitudes. Carrington found that, at the beginning of each cycle, sunspots first emerged at high latitudes. As the cycle progressed, sunspots were created

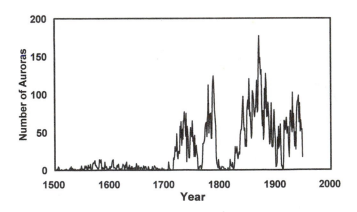

FIGURE 2.15 Secular variations in the number of aurorae (data from Silverman, 1992, with permission). The number of aurorae before the Maunder Minimum (around 1650) are probably underestimated because of lack of observations.

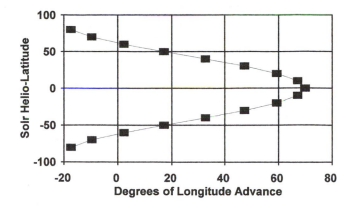

FIGURE 2.16 An example of the solar differential rotation. After 30 days or about one solar rotation, sunspots, which all initially start at zero degrees longitude, are located at the longitudes represented by the curved line. Equatorial elements have raced ahead of the polar elements. (Inspired by Stetson, 1947.)

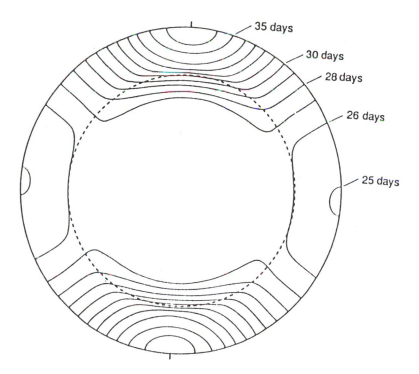

FIGURE 2.17 The interior solar rotation of the sun from helioseismology (from Big Bear Solar Observatory, California Institute of Technology, with permission). The fastest solar rotation is 25 days at the equator. The slowest rotation rates equal 35 days at each pole. Each contour is separated by 20 nanohertz or 1.48 days. Note that the surface differential rotation persists deeply into the solar interior.

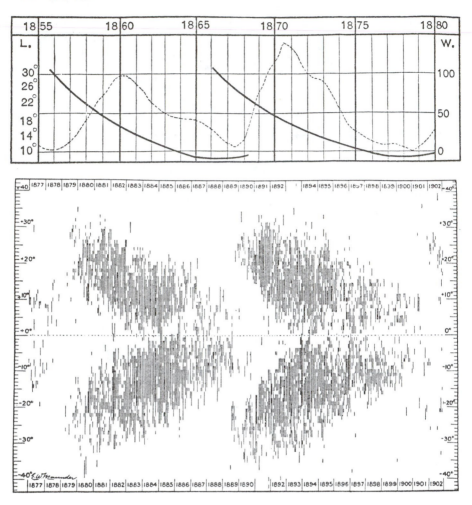

FIGURE 2.18 The Maunder Butterfly Diagram *(upper panel)* plotted in parallel with the sunspot areas *(lower panel)*. Sunspots first appear at high latitudes and, as the solar cycle progresses, are created closer and closer to the equator. The overall pattern for each cycle resembles two butterfly wings. (From David H. Hathaway, with permission.)

closer and closer to the solar equator. Gustav Spoerer in Berlin studied this phenomenon so extensively that it eventually became known as Spoerer's Law. E. Walter Maunder also studied this and in 1904 graphically illustrated its occurrence so well that ever since this type of graph has been called the Maunder Butterfly Diagram (see Figure 2.18).

Carrington's final discovery occurred in 1859 when he chanced to observe a white-light flare. Richard Hodgson also saw the flare which, on succeeding days, was followed by brilliant aurorae and magnetic disturbances. Stephen Gray also saw a white-light flare in 1705, but this earlier observation was never published.

In 1861, Carrington's father died. Soon afterward the ailing family brewery business forced Carrington to abandon his solar observations. When Carrington became a brewer, solar physics lost one of its finest observers.

George Ellery Hale and the Discovery of Magnetic Fields in Sunspots

George Ellery Hale (1908) developed the spectrohelioscope, an optical instrument capable of making detailed examinations of the sun's disk in very narrow color bands. We noted earlier that Herschel had found that portions of the solar spectrum existed outside the eye's visible range. In the following years, instruments used to view the intensity of light became more refined. J. Fraunhofer discovered in 1817 that the sun's spectrum contained sharp, dark absorption lines which arise when an electron in an atom makes a transition from a lower to a higher orbit. Only photons having precisely the proper amount of energy are involved in these transitions, causing the absorption lines to be sharp. The atom later reradiates the energy at other wavelengths. This causes the sun to emit less light at the specified color, creating a dark line in the solar spectrum. The many different elements with varying orbital levels produce the numerous absorption features in the solar spectrum.

The presence of a strong magnetic field affects the energy states of electrons in atoms. Turning his spectrohelioscope toward sunspots, Hale discovered that the absorption lines of sunspots were split in accordance with what is known as the Zeeman effect. The magnetic fields (several thousand Gauss) were many times stronger than the Earth's magnetic field (about 0.5 Gauss). These same fields also often occupy areas in excess of 10^9 square kilometers.

The magnetic fields have several additional effects. First, they inhibit heat transport into the field region and thus are associated with a cooling of the gases within the spot, which causes sunspots to be dark. A sunspot's central umbra may be only about 4200 °K compared with the surrounding solar photosphere of about 6000 °K. The typical penumbra averages about 5700 °K. Joseph Henry actually discovered this coolness of sunspots at the Smithsonian Institution in 1845, but at that time no explanation existed for it. In 1848 or 1849 Henry explained his experimental techniques to Father Angelo Secchi, who in 1852 claimed to have discovered that sunspots are cooler than the surrounding photosphere. Secchi's claim has led many astray concerning the priority of the discovery. About 1611, Battista Baliani of Genoa had speculated that sunspots were dark because they were cool, an idea that was often repeated during the following centuries. Nevertheless, no proof existed until Henry's experiments.

Hale's observations showed that earlier theories suggesting that sunspots were clouds, openings in clouds, volcanoes, or other phenomena were incorrect. Hale viewed sunspots as gigantic magnetic solar storms somewhat akin to hurricanes on Earth. This theory, although dated, may not be completely off base because, although there are obvious differences, there may also be some similarities associated with energy transport in these processes. The interior of the sunspots may be viewed as a cool, low-pressure system that causes a second effect—the depression of the sun's surface known as the Wilson Effect or Wilson Depression (see Figure 2.9).

The Solar Dynamo and Solar Activity Predictions

During the past two decades, a growing group of solar forecasters have relied on precursor methods to predict solar activity. An offshoot called the "dynamo method" has allowed the precursor method to stand on a physical foundation. Geomagnetic precursor methods started with the Soviet astrophysicists A. I. Ohl and G. I. Ohl in 1979 and the British astrophysicists G. M. Brown and W. R. Williams in 1969. Brown and Williams noticed that fluctuations in the Earth's magnetic field (geomagnetic activity) could be used to predict solar activity. Although the correlation was statistically significant, no physical mechanism existed to explain it. These scientists noted some puzzling connections: solar activity seemed to be correlated with past levels of geomagnetic activity. This was puzzling for two related reasons: (1) the sun should drive terrestrial magnetic activity and (2) the solar signal should precede, not follow, the Earth's response.

An understanding of the physical mechanisms was developed that allowed the previous correlation to be based on physical laws rather than just on statistics. We now discuss the solar dynamo, followed by the dynamo method of forecasting solar activity, and then see how these relate to the "geomagnetic precursor" methods.

The Solar Dynamo

To paraphrase Cal Tech physicist Robert Leighton, if the sun did not possess a magnetic field, it would be as uninteresting as most astronomers think it is! Thus solar magnetism, which forms through "dynamo" processes, is what makes the sun diverse and interesting. Although the solar dynamo is only generally understood, the details are both scientifically and literally shrouded in mystery due to the sun's photospheric "cloak" that prevents us from directly viewing more deeply. New probing tools, such as detecting neutrinos or using sound waves (helioseismology), are now becoming important sources of information about the solar interior.

The solar dynamo is considered to be a "migratory" dynamo, with magnification of fields occurring at different favorable locations. The sun's magnetic fields are magnified by the energy sources of convection and differential rotation. These stretching motions amplify the magnetic field. Although the details are not well known, we do know that the sun's differential rotation transforms an initially poloidal dipole field into a toroidal field (Figure 2.19). A poloidal field is a field shaped like a pole; a toroidal field is wound up in the shape of a doughnut, or toroid. Hydrodynamic forces allow the toroidal field elements to rise to the sun's surface where they erupt to form an active region. The remnants of this process, aided by flows (i.e., meridional circulation and eddy diffusion), reemerge as another poloidal field from the toroidal sunspot fields.

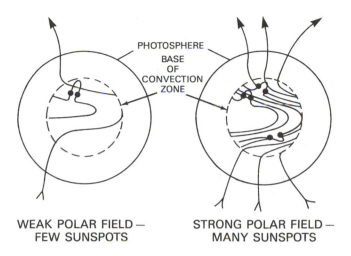

PHOTOSPHERE
BASE
OF
CONVECTION
ZONE

WEAK POLAR FIELD —
FEW SUNSPOTS

STRONG POLAR FIELD —
MANY SUNSPOTS

FIGURE 2.19 Solar dynamo. An initial weak poloidal field is wound up into a toroidal field. Toroidal elements rise to the surface to form sunspots. Eventually the poloidal field is reestablished with an opposite polarity, and the solar cycle starts anew When the polar field of the sun is weak (*left*), near solar minimum, few sunspots form and when strong (*right*), many spots appear.

The advection of magnetic fields, combined with reconnection processes, not only cancels the initial poloidal field but also creates a new poloidal field of opposite polarity. This leads to a reversal of the solar magnetic field roughly every 11 years.

Many well-known observational "laws" of solar magnetism—including Hale's law of sunspot polarities, the increase of the sun's polar field toward a dipole maximum near the solar activity minimum, and the emergent behavior of sunspots in accordance with Spoerer's butterfly diagram—appear to support the solar dynamo (Babcock model) view. While the behavior of the magnetic field in the photosphere suggests that the Babcock model is fairly complete, many unknown aspects of the solar dynamo remain. Does the field below the solar surface have a fibril or fiber-like form (most likely), or does it extend weakly over a large volume? What powers active regions? Rather than focus on these unclear aspects of the sun's dynamo, one central aspect of dynamo physics allows predictions of future solar activity—future solar magnetic fields arise directly from the magnification of preexisting magnetic fields. This phenomenon provides a degree of persistence to solar magnetism over decadal time scales. This persistence allows future levels to be predicted from current levels, even as the solar magnetism levels oscillate between toroidal and poloidal components during a solar cycle. Let us discuss the details.

Solar Activity Predictions

To begin, let us examine how future solar magnetic fields arise from preexisting magnetic fields. In the detailed Babcock model, at the start of a cycle, dynamo processes transform the sun's polar field into an internal toroidal field that, years later, increases during the maximum phase. The sun's polar field then serves as a "seed" that can be used to forecast future solar activity. Figure 2.19 illustrates the solar dynamo's field geometry, with each view showing both the polar and toroidal fields. Weak polar fields generate small toroidal fields and produce few sunspots; at the solar minimum, when the polar field is strong, the reverse is true. Figure 2.20 shows the next solar-cycle predictions for F10.7 cm Radio Flux and sunspot number based on the dynamo method. The so-called precursor methods that use geomagnetic field fluctuations at the solar minimum to predict future solar activity basically rely on the polar field relationship discussed above. At solar minimum, the solar wind sweeps the sun's polar fields past the Earth and allows the extended solar fields to influence the degree of geomagnetic field variations.

Using these views in 1978, Schatten and his colleagues developed the "dynamo method" to predict solar activity. The strength of the sun's polar fields was suggested as the precursor for future activity, and rather than rely on the geomagnetic precursors, the authors went directly to solar indices to predict future levels of solar activity.

This approach seems to work well, but further testing is clearly needed. The dynamo method was tested with eight prior solar cycles and was subsequently able to correctly predict the magnitude of solar cycles 21 and 22 several

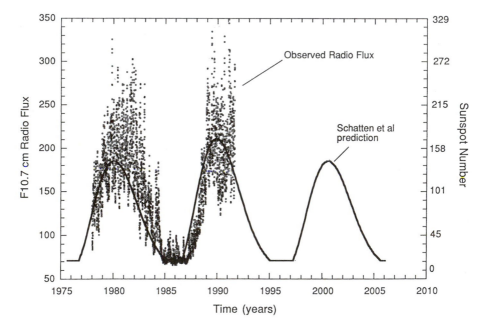

FIGURE 2.20 Previous predictions of solar activity by Schatten and colleagues in 1978 and 1987 *(solid curve),* along with subsequent Radio Flux *(left scale)* and approximate value of the sunspot number *(right scale).*

years in advance. Near the solar minimum preceding cycle 21, Schatten predicted that solar cycle 21 would peak with a sunspot number of 140 ± 20 (1 standard deviation). The subsequent peak for this cycle was 163, and the prediction was correct to about 1 standard deviation. Shortly after the next solar minimum in 1986, the dynamo method was again tested. This time it went against tradition (where even-numbered cycles are smaller than the preceding cycle) by predicting that cycle 22 would be an exceptionally large even-numbered cycle—the second largest cycle in recorded history. The dynamo method proved its worth when, several years in advance, it correctly predicted the magnitude of the cycle peak for F10.7 Radio Flux to be 210 ± 25. The actual peak was 213. Figure 2.20 shows the predictions made by Schatten et al. using this technique.

Summary

This chapter began with a historical overview of solar activity observations from their discovery to modern times. After sunspots were first discovered, two centuries elapsed before Heinrich Schwabe realized sunspots come and go in an 11-year cycle. This long delay is attributed in part to the absence of systematic sunspot observations. During regular solar observations in the 1600s, the sunspot cycle was in abeyance. Walter Maunder spent 30 years trying to attract attention to this anomaly, which today is named the Maunder Minimum in his honor.

Although we know a great deal about the sun today, much remains unknown. The mechanism of solar energy generation is well known. The solar interior consists of a radiative core surrounded by a convective envelope. Sunspots, which are generated in the convection zone, have strong magnetic fields. Modern dynamo theories have achieved some success explaining the sunspot cycle, and the strength of next sunspot cycle may be predicted based on physical rather than simply statistical methods.

Yet for all these successes, nearly an equal number of enigmas remain to be solved. Not everyone agrees with the dynamo theory. Anomalies like the Maunder Minimum are still unexplained. We do not know why solar neutrino production is lower than theories predict. Enough puzzles remain so that other possible solar behaviors could affect the Earth and humanity. One such effect could occur through changes in the sun's brightness, the subject of our next chapter.

3. *Variations in Solar Brightness*

In the last chapter we saw that sunspots, aurorae, and geomagnetic disturbances vary in an 11-year cycle. So do many other solar features, including faculae and plages, which are bright regions seen in visible and monochromatic light, respectively. If both bright faculae and dark sunspots follow 11-year cycles, does this mean the sun's total light output varies? Or are these two contrasting features balanced so that the sun's output of light remains constant? The light output of the sun is often discussed in two different ways: either as the solar luminosity, which is the sun's omnidirectional radiant output, or as the solar constant, the output seen in the direction of the Earth. In this chapter, we explore the variable solar light output that has been the subject of vigorous discussions.

The total solar irradiance or solar constant is defined as the total radiant power passing through a unit area at Earth's mean orbital distance of 1 astronomical unit. Today the most common units of solar irradiance are watts per square meter (W/m^2). Power is defined as energy per unit time, so the solar irradiance can also be expressed in calories per square centimeter per minute. Modern experiments indicate that the sun's radiant output is about 1367 W/m^2, with an uncertainty of about 4 W/m^2. About 150 years of effort by many people have been required to establish the value to this accuracy. The sun's radiant output is not an easy quantity to measure, and we will discuss some of the struggles required to measure it.

In the late 1800s, many scientists considered the solar total irradiance or solar irradiance to be constant. Oceanographers Dove and Maury vigorously supported this viewpoint, so the solar irradiance was called the solar constant. For the next century, virtually every paper concerning the sun's radiant output used the term solar constant. No physical justification for this nomenclature existed, only a philosophical bias. Yet by the 1950s this bias proved so strong and so prevalent that support for individuals who wished to measure variations in the solar constant became almost nonexistent. Only a few brave souls dared to go against the prevalent paradigm that "it is ridiculous to try to measure variations in a constant."

Herschel's Thoughts on the Variations in Solar Light

Around 1800, Sir William Herschel was a fairly active observer of sunspots and solar activity. Herschel often speculated on the meaning of what he saw, and a particular speculation interests us here. In 1801 he thought the more frequent appearance of sunspots and related activity led to the more "copious emission of heat and light." He based this conclusion on the history of wheat prices, noting that when wheat was expensive, sunspots were few. Thus, Herschel reasoned that the absence of sunspots meant the sun was emitting less light and heat, causing less wheat to grow, which, in turn, drove prices up. Herschel clarified his statements somewhat in 1807 in the *Philosophical Transactions,* noting that a change in the sun's radiant output might not have an immediate, but a delayed effect on wheat growth and prices. He concluded that "the whole theory of the symptomatic disposition of the Sun is only proposed as an experiment to be made."

C. Piazzi Smyth's Experiment

Herschel proposed that experiments should be made to measure the sun's variable output. In succeeding years, many pioneer scientists such as Bouguer, Leslie, John Herschel, Pouillet, Melloni, Soret, Forbes, Violle, Ericson, and Crova did just that. Most of these early observers attempted to make only one measurement of the solar irradiance, and their results diverged by more than a factor of two. Early instruments and analysis techniques proved inadequate to provide a definitive answer.

One early paper by C. Piazzi Smyth in 1856 entitled "Note on the Constancy of Solar Radiation" does stand out. A Scottish astronomer, Smyth is perhaps most famous today for claiming that the value of π could be found in many measurements of the Giant Pyramid at Giza. For example, he claimed that the tangent of the pyramid's slope equaled $4/\pi$. Smyth's most lasting contributions were in spectroscopy and advocating high mountain sites for astronomical observatories.

In 1837 J. D. Forbes constructed the original experiment described by Smyth. Bulbs containing alcohol were buried in porphyry rock at depths of 3, 6, 12, and 24 feet, allowing temperature measurements accurate to 0.01 °F. The deeper the thermometer, the more the measurements reveal that the annual cycles gradually become damped. Figure 3.1 shows Smyth's annual mean temperatures, along with the Wolf Sunspot Numbers.

In 1856 Smyth said the temperature variations "have at once an indication of our Sun being amongst the number of variable stars." The experiment was designed with a very slow response time to eliminate weather effects and emphasize a solar signal. Today we might argue that a change in climate, such as a decrease in mean cloud cover, is being observed here, but this explanation did not occur to Smyth, who believed the sun was undergoing a long-term increase in brightness unrelated to the cyclical variations in solar activity.

Langley's Paper of 1876

Most early papers dealing with changes in solar brightness and climate were statistical studies. Samuel Pierpont Langley's 1876 study took a decidedly physical approach that resembles modern studies in its procedure and conclusions. Born to a prominent old Roxbury, Massachusetts, family in 1834, even as a child Langley was fascinated by scientific questions and by the sun. Langley once reminisced that "one of the most wonderful things to me was the Sun, and as to how it heated the earth. I used to hold my hands up to it and wonder how the rays made them warm, and where the heat came from and how. I asked many questions, but I could get no satisfactory replies, and some of these childish questions have occupied many years of my later life in answering." Before starting his career in astronomy in 1866, Langley worked from 1851 to 1864 as a civil engineer. From 1867 to 1887 he directed the Allegheny Observatory, and during this period he developed a renewed interest concerning whether or not the sun's radiant output was constant. Langley repeated some of Henry's and Secchi's work in 1873 and 1874 to compare the heat output of solar umbrae and penumbrae with the surrounding photosphere. He found the umbra to be 0.54 times as bright as the photosphere and the penumbra 0.80 times as bright. Today's values for these same quantities are 0.24 and 0.77, respectively. Langley's high values for an umbra seem to suggest he was having

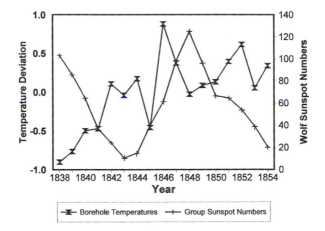

FIGURE 3.1 C. Piazzi Smyth's rock borehole temperatures from 1838 to 1854 *(triangles),* along with the Wolf Sunspot Numbers (plus signs). Smyth attributed the temperature variations to variations in solar output. These observations are the first attempt to plot the temporal variations for solar output.

FIGURE 3.2 An engraving showing the wild mountain scenery near Mt. Whitney, where Langley measured the sun's radiant output. (From Langley, 1884.)

problems with scattered light. His numbers suggest that overall sunspots block about 27% of the radiation, compared to about 37% for modern estimates. Langley then calculated that changes in sunspot blocking from sunspot maxima to sunspot minima would cause a change in solar brightness of about 0.09% to 0.10%, with the solar maxima producing a less-bright sun. The absolute magnitude of this theoretical number is close to modern measurements. Langley then deduces that the overall climatic effect would be equivalent to a change in Earth temperature between 0.063 °C and 0.29 °C. Much of Langley's work anticipates modern calculations and results, except that he did not consider the effects of excess emission by faculae.

The Mt. Whitney Expedition

Given his interest in actual solar brightness and its variations, in 1881 Langley undertook an expedition to Mt. Whitney, California, to help solve these questions (Figure 3.2). This expedition is interesting both in its own right and because of the difficulties encountered when making measurements of this type.

To get above most of the Earth's atmosphere, which complicates any ground-based measurement, Langley sought to measure the solar irradiance from a high mountain site. After considering several sites, he chose California's Mt. Whitney, which rises more than 14,000 feet (4200 meters) above sea level and above about 40% of the atmosphere. High-altitude measurements are needed because the atmosphere both absorbs and scatters radiation. If the atmosphere remains unchanging during the day, these effects can be removed at a single wavelength. At many wavelengths, the measured intensity at the Earth's surface equals the extraterrestrial solar irradiance diminished by atmospheric scattering alone. The equation for this variation is

$$I_{meas} = I_0 e^{-\tau^* m} \tag{1}$$

where I_{meas} is the measured irradiance, I_0 is the extraterrestrial irradiance, τ is the optical depth, and m is the air mass, which is approximately equal to 1 over the cosine of the solar zenith angle, or $1/\cos(\theta)$. The air mass is the amount of atmosphere between the observer and the sun. Looking straight up at sea level, the air mass is equal to 1. At sea level with the sun at 60° from the zenith, the air mass equals about 2. For sea-level sites, the air mass equals the secant of the solar zenith angle over a fairly wide range of angles. If one could get above the atmosphere, the air mass would equal zero and I_{meas} would equal I_0. The optical depth measures light attenuation in a vertical path, mostly caused by scattering by air molecules or dust particles. τ is a dimensionless quantity that, for the Earth's atmosphere, typically has values of a few tenths. Its values are larger for shorter wavelengths. In general, the optical depth can be found by solving the equation above as follows:

$$\tau^* m = \ln I_0 - \ln I_{meas} \tag{2}$$

Thus, if one calculates m and measures I_{meas} during the course of a morning, then a plot of $\ln I_{meas}$ versus air mass will be a straight line. The slope will equal the optical depth, and the intercept will equal $\ln I_0$. At each wavelength one can then get the extraterrestrial solar irradiance. Today, these are called Langley plots.

Of course, carrying out this simple procedure proves to be difficult. For example, if the atmosphere's transmission changes during the day, the straight line will become curved and the intercept will be meaningless. If absorption occurs along with scattering for a selected wavelength, the results will no longer be valid, in part because atmospheric absorbers such as water vapor are not uniformly mixed. Finally, to work properly the irradiance measurements must have a reliable radiation scale. All these and other problems would have hindered Langley.

Given financial support by the Signal Service and William Shaw, Langley hastily organized the expedition without checking all his instruments before their shipment to California. His haste proved costly. On July 7, 1881, Langley, accompanied by three assistants, left by train for Caliente, California. From there they traveled 70 miles by wagon to Lone Pine at the base of Mt. Whitney, where they established the lower observatory. Getting the instruments to Lone Pine and to the peak of Mt. Whitney proved to be a formidable problem. While crossing the 70 miles of desert between Caliente and Lone Pine, dust entered every box containing the delicate optical instruments. In his haste, Langley had arrived in California without all his expected instruments, including an acti-nometer built by Very to measure the intensity of the sun's disk (I_{meas} in equation 1) which never arrived. Today actinometers are called pyrheliometers. The Marie-Davy actinometer was broken before its arrival at Allegheny and, as of July 30th, another actinometer built by Crova in France had still not arrived. Nevertheless, Langley and his expedition departed the lower camp at Lone Pine by mule train, carrying what instruments they had to the upper camp. Only a single actinometer arrived in working condition at the upper camp, with many other valuable instruments left broken and abandoned by the mule drivers be-side the trail. The instrument designed to view the solar corona could not be made to work, and the spectrobolometer only became operable on August 31st. Then on September 4th, the now famous California wildfires began making the sky smoky and hazy. At about this time, a critical rock salt prism was left out overnight and was damaged by moisture. Then the newly constructed 5-inch equatorial telescope was found to be incompatible with the eyepieces borrowed from another telescope. By September 8th, the sky conditions had worsened to such an extent that they threatened the whole experiment, and the next day Langley began descending the mountain.

Despite these many difficulties, enough measurements were made to derive a solar-irradiance value. In 1884 the expedition's results were published in the Professional Papers of the Signal Service as a 239-page monograph called *Researches on Solar Heat and Its Absorption by the Earth's Atmosphere.* Langley had used the spectrobolometer to measure the solar spectrum. Figure 3.3 illustrates today's solar spectrum. Langley was not the first to measure the extraterrestrial solar spectrum, but his was the best effort up to that time. Langley was less successful obtaining an accurate value of the solar total irradiance. He

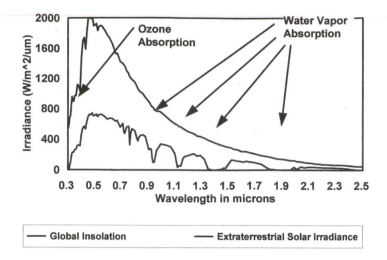

FIGURE 3.3 The extraterrestrial solar spectrum at the top of the Earth's atmosphere is plotted in the upper curve. The sun is brightest at about 5000 Ångstroms or 0.5 mircron. The lower curve shows a typical plot of the global insolation from the sun and sky at the Earth's surface. It has strong absorption bands in the infrared caused by water vapor and is attenuated in the ultraviolet by ozone absorption. The curve has a very similar shape to Langley's first determination of the extraterrestrial solar irradiance. Values for the extraterrestrial spectrum are tabulated in Appendix 2.

deduced a value of 2903 W/m^2 (or 3 cal/cm^2 per min), a high value attributable in part to mathematical errors in the data reduction. Using the same data, Abbot (then Langley's assistant) later found that the expedition's correct number should have been 1465 W/m^2 (or 2.1 cal/cm^2 per min), very close to our modern value.

The APO Solar-Constant Program, 1902 to 1957— Or How Does the Solar Light Vary?

In June 1895, 23-year-old Charles Greeley Abbot was working in his laboratory at the Massachusetts Institute of Technology (MIT) toward a master's degree in physics. Suddenly he was interrupted by the message, "Professor Langley wants to see you upstairs." Rushing upstairs in his acid-spattered overalls and a jumper, Abbot met Langley, then director of the Smithsonian Institution, for the first time. In contrast to Abbot, Langley wore a suit and silk top hat. The meeting occurred because Professor Cross of MIT had mentioned Abbot to Langley. Abbot invited Langley to see his laboratory, but Langley was in a rush and could not spare the time.

The next day Abbot received an unexpected telegram offering him a $1,200-per-year job at the Smithsonian Institution. This serendipitous event altered Abbot's life, the direction of research at the Smithsonian Institution, and

the field of sun/climate research over the next 60 years. Abbot accepted imme-
diately, began packing that same day, and took the night train to Washington.
He need not have rushed. Since Langley had left on an extended European
trip, Abbot spent the summer in the rickety building housing the Smithsonian
Astrophysical Observatory (APO) with nothing to do except to endure the swel-
tering 120° heat inside. Robert Child, Frederic E. Fowle, and Charles Greeley
Abbot comprised the observatory staff, and upon his return in the autumn,
Langley first and most warmly greeted Abbot. In time, under Langley's mentor-
ship, Abbot became head of the APO and, following Langley's death in 1906,
he became secretary of the Smithsonian Institution. Abbot was to remain with
the Smithsonian until 1954. From 1902 to 1957 the APO made thousands of
determinations of the solar irradiance, the most extended experimental cam-
paign ever mounted to search for changes in solar brightness.

When the APO began its measurement program, the solar-irradiance value
was poorly known, with published values ranging from 1230 to 2790 W/m^2.
Whether the sun's light output was constant or variable had not been determined. In 1895 Lockyer had shown that some solar absorption line variations
occurred and that these variations might indicate a change in the solar output.
Other, more indirect, arguments attributed climate changes to solar changes. In
1902 Abbot stated that "there seems to be a preponderance of suggestion that
the Sun radiates more at sunspot maximum, although there are not wanting
many who hold precisely the contrary opinion." If Abbot had any prejudice at
all before starting his measurements, he seemed inclined toward the belief that
a more active sun implied a brighter sun. However, Langley stated "my personal bias, so far as I have any, would incline me to wish to see a change in
the solar constant established. The possibility of such a bias existing is, however, only a reason to me for additional caution."

By 1902, Abbot and the APO had developed a whole suite of instruments
to measure the solar irradiance, including water-flow pyrheliometers to fix the
absolute values of the radiation, silver-disk pyrheliometers to measure the sun's
brightness at the Earth's surface, and spectrobolometers to measure the intensity
of the solar spectrum. The APO began its measurements in hazy downtown
Washington, D.C., and in less than a year seems to have detected a 10% decrease in the solar brightness. Unfortunately, the 1902 volcanic eruptions of
Santa Maria, Soufriere, and Mt. Pelee had all injected large quantities of aerosols into the stratosphere, creating faulty measurements. These volcanic aerosols spread north and eventually decreased the atmospheric transmission over
Washington, D.C. Nonetheless, Abbot thought he had successfully detected a
change in the sun. Instead, his corrections to remove atmospheric effects were
simply inadequate. Coupled with an apparent worldwide decrease in temperatures, this seeming change in the sun and climate nonetheless proved a major
motivation to continue measurements. If the APO had not "detected" a solar
change, measurements would likely have soon ceased. In 1910 Abbot wrote,
"It is not probable that I should have been here this evening if it had not
happened that our 'solar constant' values of 1903 suggested a fall of solar
radiation of about 10 percent at a time just before there occurred a general fall

of several degrees centigrade from the normal temperatures of the United States and Europe."

Atmospheric contamination of the measurements continued to be problematic for the 55 years from 1902 to 1957. To avoid this problem, Abbot began establishing remote observing sites on mountains to get above the atmosphere. His first site was established in 1905 on Mt. Wilson in California, and in time he would establish sites at Mt. Whitney (Calif.), Hump Mountain (N.C.), Calama (Chile), Mt. Harqua Harla (Ariz.), Mt. Montezuma (Chile), Table Mountain (Calif.), Mt. Brukkaros (South Africa), Mt. St. Katherine (Egypt), Tyrone, (N.M.), and Miami (Fla.). The greatest success was achieved after 1923 at Mt. Montezuma and Table Mountain. By then, techniques existed to remove atmospheric effects, the sites were stable and well established, and the instruments were fully developed.

The findings of Abbot and the APO differ depending on who analyzes the data. Let us start with what Abbot thought he found. H. H. Clayton was a meteorologist and active supporter of Abbot. Abbot's book *The Sun and the Welfare of Man* states:

> H. H. Clayton has numerically studied the value of the solar constant as sunspots cross the center of the Sun's disk in the course of the solar rotation. For the mean of 36 of the larger spots, occurring between July 1918 and July 1924, he finds that the solar radiation decreased by three-tenths per cent on the day following their central passage. However, not all sun-spots are thus attended by a depressed solar radiation, any more than all sun-spots are attended by northern lights or magnetic storms. Hence the average depression found by Clayton for all large spots is less than the average depression attending active spots only.

From this quotation we conclude that Abbot believed these measurements contained evidence of sunspot blocking. It seems reasonable that this type of variation, confirmed by modern satellite observations, did exist in his observations.

Along with these short-term variations, Abbot concluded that he had also found long-term variations in solar irradiance, with a more active sun being brighter. Figure 3.4 shows the dependence on solar activity deduced by Abbot. However, problems exist with his conclusions. The first problem is Abbot's claim that the active sun is about 1% brighter than the quiet sun. The magnitude of this change is not substantiated by modern observations; however, the direction is. The second problem is that others who have analyzed Abbot's data do not arrive at the same conclusions. In 1979 we examined Abbot's monthly mean observations from 1923 to 1954 and found the solar-irradiance (S) dependence on Wolf Sunspot Number (R_z) to be:

$$S = (1357.2 \pm 2.86) + (0.00488 \pm 0.00419)R_z \qquad (3)$$

Although the sign and size of this dependence agree with modern observations, the results have such large error bars, the uncertainties make the findings not statistically significant.

Using Abbot's daily observations, in 1977 P. V. Foukal and his colleagues found that higher values of solar irradiance were associated with more solar

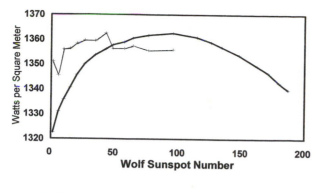

FIGURE 3.4 The dependence of solar total irradiance on solar activity, using the Wolf Sunspot Numbers, as deduced by Abbot (data from Abbot, 1934) and again by Kondratyev based on measurements from 1962 to 1967 (data from Kondratyev and Nikolsky, 1970). Abbot's solar irradiance varies about 1%, but Kondratyev's variation is 3%.

faculae. Their conclusion that the measurements showed facular emission was an important contributor to solar-irradiance variations and in accord with modern thinking.

Although at least three correct conclusions concerning solar behavior were deduced using Abbot's measurements, Abbot still had his critics. For example, in an article entitled "The constancy of the solar constant," T. E. Sterne and N. Dieter argued that the APO measurements showed that the solar output over many years was constant to 0.1% or better. This upper limit on solar variations was based on the assumption that the APO measurements made at different sites were completely independent. Unfortunately, this assumption is not true. Abbot and his colleagues adjusted the different stations to agree better with each other. Hence, the correct upper limit is about 0.3% rather than 0.1%. Variations of this amount are still significant, both climatically and scientifically. Because of the variability in atmospheric transmission, it is unlikely that any ground-based solar-irradiance observation could put stricter limits on solar variability. Better measurements require observing above the atmosphere.

Abbot's efforts produced both successes and failures. His successes include getting the correct sign and sometimes the correct amount for the variations in solar irradiance. His failures include being unable to convince the scientific community of the correctness of his results, primarily because of the limited accuracy and repeatability of his measurements. Above and beyond any success or failure in the realm of solar physics, Abbot and his colleagues left behind a remarkable data set that has proved useful in monitoring changes in the composition and transparency of the Earth's atmosphere. These measurements have been used to detect the injection of volcanic aerosols into the stratosphere and to detect volcanic eruptions overlooked by others. They have shown that, excluding changes in atmospheric transmission due to volcanoes, most locations have remarkably constant transmissions for 30 years or more. Some of Abbot's measurements have also been used to deduce the concentration of atmospheric carbon dioxide, and his data are valuable to anyone interested in secular variations in the Earth's atmosphere and how these secular variations may affect climate change.

What Is the Mean Solar Irradiance?

While Langley and Abbot were busy measuring the solar irradiance, many others were performing the same types of experiments. Rather than review all these efforts, the results are briefly summarized in Table 3.1.

Today the solar irradiance is believed to be about 1367 W/m^2, but its true value remains uncertain by about 4 W/m^2.

Three Satellite Measurements of the Solar Total Irradiance: Nimbus-7, SMM/ACRIM, and ERBE

Before the launch of satellites with stable radiometers, scientists supported four different positions concerning the sun. The most popular opinion by far was that the solar irradiance was constant and unvarying. Climatologists who could find no persuasive evidence that a changing sun was affecting their studies strongly supported this belief. The second most popular opinion was that the sun was brighter when the sun was active. Abbot and many other scientists supported this idea. The third most popular idea was that the sun was less bright when dominated by active sunspot blocking. Finally, a very small number of scientists believed the sun's brightness was changing in a way not related to solar activity, a theory C. Piazzi Smyth proposed in 1856.

Ground-based, balloon, and rocket flights were all problematic enough to make any answers they provided too uncertain to settle the issue. Only a high-accuracy, high-stability satellite-borne radiometer would do. By the mid-1970s, several groups of scientists were proposing that an electrically calibrated cavity radiometer be placed aboard a satellite to measure the solar total irradiance. Radiometers of this design had a black painted cavity that absorbed nearly all the solar radiation falling on it. The absorbed radiation raised the temperature of the cavity so that a radiant power could be measured corresponding to the rise in temperature. Each cavity could also be heated by an electrical element in a manner nearly equivalent to the incident sunlight. Since the injected electrical power can be measured accurately, the radiometer's temperature response can be calibrated. Radiometers of this design are all self-calibrating.

In the 1970s several radiometers of this type existed. At the Jet Propulsion Laboratory in Pasadena, California, Jim Kendall Sr. had a radiometer called the Primary Absolute Cavity Radiometer (PACRAD), the first self-calibrating radiometer to be launched into space on board the Mariner 2 mission to Mars. Although its measurements suggested that small changes might be occurring in the sun's radiant output, the results were not intensively investigated.

TABLE 3.1 A few of the many measurements of the solar total irradiance. Values are expressed in modern units.

Author	Observation or Publication Date	Solar Total Irradiance (W/m^2)
Pouillet	1838	1230
Forbes	1842	1988
Herschel	1847	1458
Crova	1875	1324
Violle	1879	1772
Langley	1884	2903
Abbot	1904	1465
Abbot	1923–1954	1358
Linke	1932	1354
Mulders	1934–1935	1361
Unsold	1938	1326
Moon	1940	1322
Aldrich & Abbot	1948	1325
Schuepp	1949	1367–1416
Allen	1950	1374
Nicolet	1951	1382
Aldrich & Hoover	1952	1349
Johnson	1954	1395
Sitnik	1967	1448
Drummond	1968	1360
Duncan & Webb	1968	1349
Kruger	1968	1358
McNutt & Riley	1968	1343–1362
Stair & Ellis	1968	1360–1370
VonderHaar	1968	1390
Arvenson et al.	1968	1355–1365
JPL–Mariner 6 & 7	1969	1355
Murcray et al.	1969	1338
Thekaekara et al.	1969	1352
Kondratyev & Nikolsky	1970	1353
Labs & Neckel	1970	1358
Willson	1971	1370
Nimbus-7	1978–1993	1372
SMM/ACRIM	1980–1988	1368
ERBE	1984–1993	1365
UARS/ACRIM	1993–	1365

Another cavity radiometer built by John R. Hickey at the Eppley Laboratory in Newport, Rhode Island, and known as the Hickey-Frieden radiometer, originally had no assigned satellite mission. Hickey and his colleagues had previously tried to measure the solar irradiance on the Nimbus-6 satellite using a flat plate thermopile radiometer. Radiometers of this design are not self-calibrating and not stable because they have flat plate detectors. The harsh conditions in space, with various forms of harsh radiation, free oxygen radicals, and so on, conspire to change the properties of radiometers. The paint on these

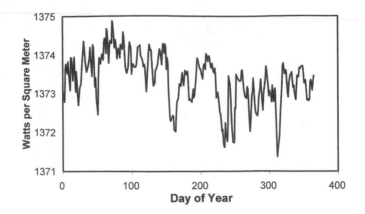

FIGURE 3.5 The solar total irradiance measured by Nimbus-7 during 1979, its first full calendar year of operation. John Hickey attributed the large dip in August to sunspot blocking. (Adapted from Hoyt et al., 1992.)

early radiometers rapidly became less black, so less radiant energy was absorbed, and their output steadily decreased. Extracting these instrumental changes and recovering the actual solar behavior is virtually impossible. NASA was planning to replace Nimbus-6 by launching Nimbus-7, which used the same flat plate radiometers, but John Hickey had other ideas. Removing the flat plate radiometer on short notice, he replaced it with the Hickey-Frieden cavity radiometer. The Nimbus-7 satellite was launched in 1978 with its unauthorized payload, and on November 16 the cavity radiometer began acquiring useful data.

The Hickey-Frieden cavity radiometer's earliest measurements clearly proved the sun to be a variable star. Figure 3.5 shows the first full calendar year of the Nimbus-7 solar irradiances. Hickey attributed the large dips in the measurements, such as the one in August, to sunspot blocking. This figure clearly shows the sunspot signal, which appeared much noisier when the data were first examined. Some "noise" was due to variations introduced because the radiometer was not pointed at the sun but about 2.4° away. An apparent jump in solar irradiance appears each time the pointing was adjusted by 1°. With the instrument's limited sampling and resolution, the radiometer signal was initially assumed simply to be noisy. This problem was identified and corrected 10 years later when the data were examined by a person new to the project.

In 1980 the Solar Maximum Mission was launched with another cavity radiometer. Built by Richard C. Willson of the Jet Propulsion Laboratory, this radiometer was called the Activity Cavity Radiometer, or ACRIM. The ACRIM has a blackened cavity with electrical wires wound around it. Electrical heating kept the cavity at a constant temperature. The cavity is alternately exposed and shielded from the sun. The amount of electrical heat required to keep the cavity at a constant temperature is measured, and the differences in electrical heating are equal to the differences in radiant solar heating. This electrical substitution method permits very accurate measurement of the radiant heat. The ACRIM had several advantages over the Nimbus-7 instrument. First, a better analog-to-digital convertor provided better resolution. Second, it measured the sun for about 45 minutes during each orbit, or 12 hours per day, compared to 13 or 14 times by Nimbus-7. Third, two backup ACRIMs were available that were infrequently exposed. Comparisons of the three ACRIMs allowed Willson to monitor possible degradation in his instrument. Because of these superior features, ACRIM obtained a very quiet signal, and observing events, such as sunspot blocking, proved easy. In 1981 Willson and colleagues published an article containing a figure showing the sunspot blocking along with images of a large sunspot group crossing the solar disk (see Figure 3.6). The ACRIM measurements give the first clear indications of excess emission by faculae (bright solar features in active regions). Surrounding the dips due to sunspots are periods of excess solar brightness caused by faculae.

The ERBE (Earth Radiation Budget Experiment) satellites included a third series of cavity radiometers. Three satellites (ERBS, NOAA-9, and NOAA-10) had activity cavity radiometers primarily to aid in the calibration of the Earth-viewing radiometers. A secondary objective was measurement of the solar irradiance. These measurements were made only once every 12 days and so do not resolve such features as large sunspot groups. Nevertheless, they can resolve variations in the solar irradiance on the time scale of a solar cycle. Robert Lee of NASA's Langley Research Center runs this measurement program.

Figure 3.7 shows the daily measurements from three satellites carrying instruments to measure the solar irradiance from 1978 to 1993. The greater variability in the solar irradiance when the sun is more active is a real effect caused by sunspot blocking. Each series of measurements indicates the existence of an 11-year solar-irradiance cycle with an amplitude of about 0.15% using yearly means. Monthly means can differ by about 0.25%, and daily means can differ by up to about 0.5%. To show that solar activity and solar-irradiance values parallel each other, the monthly Nimbus-7 solar irradiances and the Wolf Sunspot Numbers are plotted in Figure 3.8. Although some differences in shape exist between the curves, notably in 1979, agreement is sufficient to suggest an 11-year solar-irradiance cycle. Figure 3.9 plots these solar irradiances versus the Wolf Sunspot Number. The dependence of solar irradiance on solar activity appears similar to that deduced by Abbot many years ago, but with considerably less amplitude variation. The dependence shows that facular emission dominates over sunspot blocking during the 11-year cycle. The equation relating these two variables is:

$$S = 1371.32 + (0.00734 \pm 0.00069)R_z \qquad (4)$$

This equation is similar to the one derived from Abbot's measurements from 1923 to 1954, but here the results are statistically significant.

Another way to look at the measurements is to plot a histogram of their values. The asymmetry in the histogram in Figure 3.10 indicates that emission features are longer lived than sunspots. Extremely low solar irradiances can

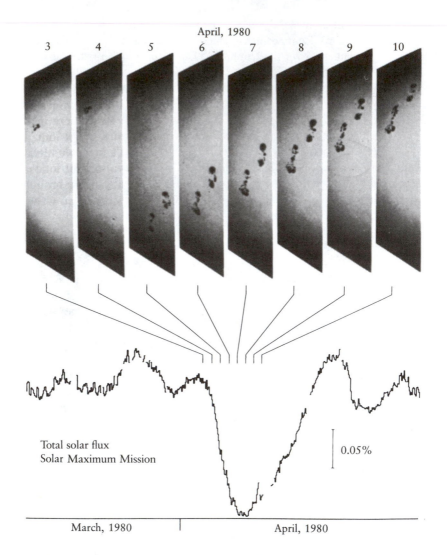

FIGURE 3.6 An example of the effects of sunspot blocking on the solar irradiance (from Willson, 1981). The upper panel shows a large sunspot group crossing the solar disk in April 1980. Each image represents 1 day. The dotted lines connect the solar image to the irradiance variations plotted below. When the spot is at the center of the disk and presents its maximum size, the solar irradiance reaches a minimum. Faculae surround sunspots but have low contrast when near the center of the sun. As the active region approaches the solar limb, facular contrast and emission increase. After the sunspot passes off the solar disk, some faculae remain visible. These faculae cause the "shoulders" or excessive emission on either side of the sunspot blocking event. (Figure from Hugh S. Hudson, with permission.)

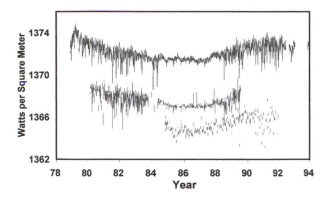

FIGURE 3.7 Solar total irradiances from three satellites in the 1980s. The offsets between the curves arise from residual uncertainties in the absolute calibration of the radiometers. (Adapted from Hoyt et al., 1992.)

occur when a large sunspot crosses the solar disk and the sun is quiet with few faculae.

Reviewing a few highlights thus far: Around 1611 Baliani first guessed the existence of sunspot blocking. In 1845 Henry first measured sunspot blocking, which was confirmed by A. Seechi in 1852. S. P. Langley first calculated its effects in 1876. In 1932 H. H. Clayton deduced the amplitude of sunspot blocking using data derived from C. G. Abbot's 1910 measurements. At a scientific

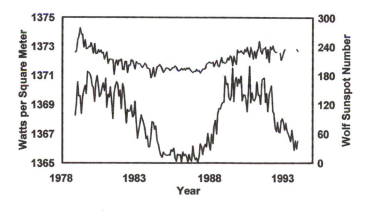

—— Solar Irradiance = Upper Curve   —— Wolf Numbers = Lower Curve

FIGURE 3.8 The monthly mean solar irradiances from Nimbus-7 and the Wolf monthly mean sunspot numbers. These two numbers parallel each other fairly well. The monthly mean solar irradiances span a range of about 0.25%. (Adapted from Hoyt et al., 1992.)

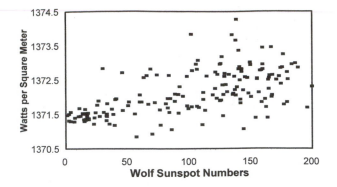

FIGURE 3.9 A scatter plot of the monthly mean solar irradiances from Nimbus-7 versus the Wolf Sunspot Numbers derived from replotting Figure 3.8. Faculae dominate over sunspots, so increased solar activity leads a brighter sun. Earlier conclusions by Abbot and Kondratyev (Figure 3.4) have the correct direction, but their variations range from 1% to 3% rather than 0.25%.

conference in Toronto in early 1980, J. R. Hickey and colleagues finally demonstrated sunspot blocking existed, and in 1980 R. C. Willson et al. confirmed this with improved data from satellite observations. Sunspot blocking appears to dominate the solar irradiance changes; however, facular emission is now known to dominate the sunspot deficits. This occurs because as the sunspot crosses the central meridian the signal is a sharp depression. The faculae, on the other hand, are more diffuse, and their enhanced output is more omnidirectional. It appears that facular outputs dominate over sunspots through a general enhancement of the solar radiative output.

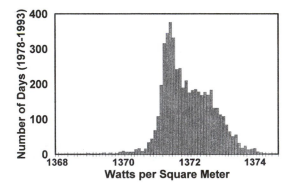

FIGURE 3.10 A histogram of Nimbus-7 daily mean solar irradiances from November 16, 1978, to July 31, 1991 (from Hoyt et al., 1992). Bin widths are 0.1 W/m^2.

Models of the Eleven-Year Solar-Irradiance Variations

Given what we now know about solar-irradiance variations and the availability of various solar indices, we can model the short-term variations. Many such models have been developed using a number of different solar indices. These are regression models, with the solar irradiance found to be dependent on sunspot blocking, facular emission, and the active network emissions. Those models are phenomenological, as opposed to physically based upon fundamental radiative principles. Nonetheless, these statistical models provide an important first step and are fairly accurate in predicting short-term variations. Long-term solar variations are much more uncertain.

The solar-irradiance models start with features that contrast with the quiet solar disk. There are dark sunspots and bright faculae, and more subtle, bright features such as the active network may be included. We have measurements of sunspot areas. Occasionally, facular areas are measured; most often, however, the more easily discernible plage areas must be used instead. Such substitutions are referred to as proxy indices for the quantity of interest. For modeling solar irradiance in the remote past, the physical causes are uncertain; nevertheless, the Wolf Sunspot Numbers can be used with some success as a substitute for the facular influence; however, there is no guarantee that a one-to-one relationship exists between sunspots and faculae. The active network does not generally have area measurements, so a proxy or substitute such as the strength of the He I 10830 Å line is used. Sunspot areas, facular areas, and active network areas (or their substitutes) can be used as the dependent variables regressed against the daily values of solar irradiances. The Foukal and Lean model (Figure 3.11) shows these impressive results. This model tracks the measurements very well, explaining about 90% of the variations. On these timescales the published literature contains many similar models with different approaches and additional refinements.

Since this model has great success describing the daily variations over many months, it is natural to attempt to extend the model over years and decades. The lower panel in Figure 3.11 shows how the last solar cycle was modeled. Although most measurements and the model track each other well, the model fails to show the large peak in 1979 that is apparent in the Nimbus-7 observations. The measurements may contain some error, so this alone would not indicate an unsuccessful model. Looking at the ACRIM measurements on the Solar Maximum Mission in early 1980, however, shows these measurements diverging from the model in the same way as the Nimbus-7 measurements. Two such measurements diverging from the model suggest that the model may be in trouble. In addition to the three components of variability already identified, there could be other as yet unidentified components causing additional solar-irradiance variability.

Despite these concerns, creating models of the solar-irradiance variability over several decades is worthwhile. These models are interesting in their own right and can be used by climatologists. Several such models exist, but here we

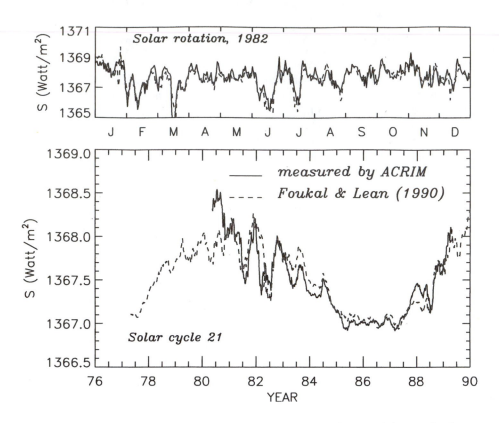

FIGURE 3.11 An example of a model for the daily solar irradiance and the actual solar irradiance measurements (from Foukal and Lean, 1990, p. 520, with permission). Many short-term variations are well modeled, but some discrepancy exists between the model and measurements near the peaks of the solar cycles *(lower panel)*.

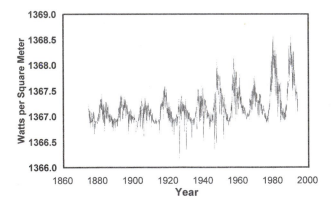

FIGURE 3.12 Foukal and Lean's model for the solar irradiance, which shows how the solar irradiance may have behaved from 1874 to 1988. In this model, the last two solar cycles have some of the largest irradiance variations ever seen. (Model data supplied by Judith Lean.)

will mention just the Foukal and Lean model shown in Figure 3.12. This model, which covers the period from 1874 to 1988, uses sunspot areas from the Royal Greenwich Observatory and substitutes the Wolf Sunspot Numbers for faculae. The model is noteworthy because the latter two cycles are the brightest seen in the last century.

Solar-Irradiance Variations as a Function of Wavelength

The variations in the sun's light-output vary with wavelength. At shorter wavelengths, the fractional variation in the solar spectrum becomes larger. For example, in the late 1920s and early 1930s Pettit measured the ultraviolet variations and found them to be much larger than Abbot's total irradiance variations. As a first approximation of this process, consider the sun as a blackbody with a temperature of 5770 °K. If, for some reason, the sun's blackbody temperature were to increase by 1 °K, so that its total output increases by 0.07%, the changes in spectral output will be greater than the mean variation for all wavelengths less than 0.6 micron (see Figure 3.13). Due to the nature of the Planck radiation law, the changes are greater at the shorter wavelengths.

The sun's output is more complicated than that of a blackbody, but the general rule of greater ultraviolet variability holds (see Figure 3.14). In the x-ray region of the solar spectrum, emission lines dominate and vary in brightness by a factor of up to 100 over a solar cycle. For some near infrared wavelengths, sunspot blocking may so dominate these wavelengths at the solar maximum that the sun will actually be dimmer. Although fractional variations are greater for shorter wavelengths, the absolute variations are equal to the product of the fractional variations and the spectral irradiance. Thus the greatest spectral irradiance variations occur near 0.5 micron, the peak of the visible spectrum.

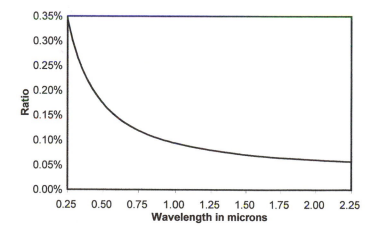

FIGURE 3.13 A plot of the ratio of 5771 to 5770 °K blackbodies. A 1-degree change in the temperature of a 5770 °K blackbody causes the total output to increase by 0.07%. For wavelengths less than about 0.6 micron the variations are greater than 0.07%, reaching more than twice the mean at about 0.26 micron.

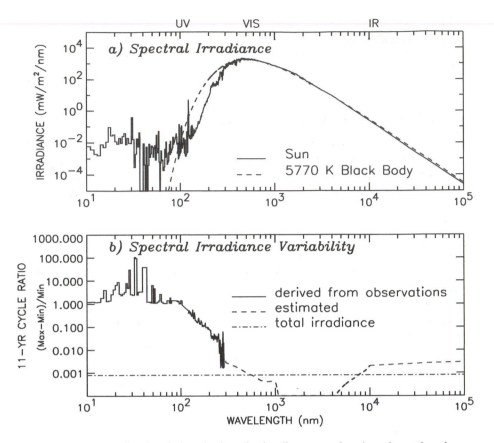

FIGURE 3.14 Fractional variations in the solar irradiance as a function of wavelength (from Lean, 1991, with permission). Larger variations occur at shorter wavelengths. Although fractional variations are greater for shorter wavelengths, the absolute variations, equal to the product of the curves in the two panels, has a shape like the curves in the upper panel. Thus, the greatest flux variations occur near 0.5 micron, the peak of the visible spectrum.

These variations complicate the sun/climate connection. Shorter wavelength radiation tends be absorbed higher in the Earth's atmosphere. At very high altitudes, solar flux variations play a dominant role in the temperature of the atmosphere (e.g., the Earth's thermosphere). Most radiation below 0.3 micron is absorbed before it enters the Earth's troposphere. Most of the spectral irradiance variation occurs around 0.5 micron and most of this radiation is absorbed after penetrating the Earth's surface. At solar maximum, the sun heats both the Earth's upper atmosphere and the surface. We will discuss the vertical distribution of this atmospheric forcing later.

Other Experimental Techniques to Measure the Sun's Light Output

Besides the direct measurement of the solar total irradiance acquired from satellites, other proxy indicators may allow the solar output to be measured from the ground. Foremost among these techniques is the examination of absorption lines in the solar spectrum. Light emerging from within the sun is continuously emitted and absorbed by atoms. Temperatures vary with height in the sun's atmosphere. Upon emerging from the sun, the light temperature first drops to a minimum value near 4500 °K, then levels off and slowly rises, followed by a rapid rise in the transition region (known as the chromosphere) to the several-million-degree solar corona. This temperature structure is illustrated in Figure 3.15.

At each temperature, probabilities exist that any atom will achieve a particular excited state. Absorption line shapes are formed at different levels in the solar atmosphere, with the core of the line forming at the temperature where the transition probabilities of an electron moving from one orbital level to an-

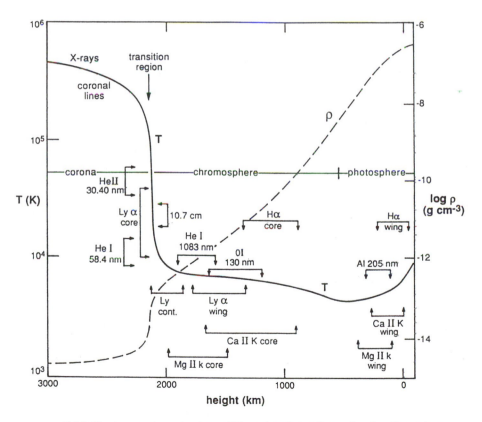

FIGURE 3.15 The temperature structure of the solar photosphere, showing the regions for formation of different absorption lines. (From W. F. Livingston, with permission.)

other are maximized corresponding to that transition, and the line's wings form at different temperature levels. Each absorption line has a preferred formation region in the solar photosphere. Those lines that absorb very little radiation are known as weak lines. These weak lines can form in very narrow layers of the sun's atmosphere, and by examining different weak lines one can explore different heights in the solar photosphere (see Figure 3.15).

Just before the turn of the century Sir Norman Lockyer announced that his measurements over the course of the solar cycle suggested changes in the strength of absorption lines. Lockyer concluded this meant that the solar total irradiance was also changing, with the sun being brighter at solar maximum. Accurate measurements of this nature are very difficult to make, so it is not clear that Lockyer's equipment was good enough to detect the effects he claimed. Since the mid-1970s, Bill Livingston at the Kitt Peak Observatory has monitored many of these lines. Using better instruments than previously available, Livingston has measured the equivalent widths of dozens of lines formed at many different depths in the solar atmosphere. The equivalent width is simply the width the line would have if it absorbed 100% of the light and had a rectangular shape. Figure 3.16 shows some of Livingston's results for 1976 to 1992. Some weak iron lines display almost no evidence of a solar cycle, but do reveal apparent long-term trends. These trends suggest that the temperature structure of the solar atmosphere may be changing over long intervals in a manner unrelated to solar activity. In turn, the changes in temperature structure imply ongoing changes in total solar luminosity, both on the 11-year time scale and a longer time scale. In addition to changes in solar irradiance being caused by sunspots, faculae, and the active network, there may be another variational component not yet detected by satellites. The possibility that these variations are real is so important that we will devote all of chapter 10 to them.

The Missing Neutrinos

Around 1970, scientists began measuring the sun's neutrino flux. Neutrinos are massless (or quite possibly low mass) particles produced during fusion in the sun's core. These particles interact very weakly with matter. They can escape the solar core directly in a couple of seconds, wheareas photons may take millions of years to make the same journey because photons are continually absorbed and reemitted. At Earth's orbit, the value for the neutrino power flux is about 33 ± 3 W/m^2. Thus, neutrinos carry energy away from the solar core. Measurement of their flux provides a window into the workings of the solar core. Using neutrinos rather than photons permits direct viewing of the sun's core. The measurements of the solar neutrinos reveal two mysteries: first, there are fewer neutrinos than predicted and, second, some evidence suggests that the flux may vary in an 11-year cycle. The first mystery suggests that either our view of stellar structure is wrong and the nuclear fusion process is either not understood correctly or there is something wrong with our understanding of fundamental physics. The implications of the second mystery are uncertain, but suggest either that the sun's core fusion rate varies with time or that earlier

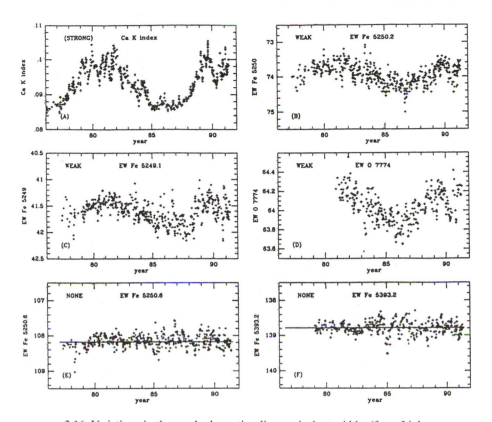

FIGURE 3.16 Variations in the weak absorption line equivalent widths (from Living-ston, 1990 with permission). These variations arise from changes in the temperature of the photosphere. Some lines do not parallel the solar cycle and suggest there may be another long-term component for solar-irradiance variations that is not yet modeled.

measurements were simply not stable and reliable. Good arguments exist that the solar fusion rate is constant and well understood. Equally good arguments counter that the stellar structure models are essentially correct, too. In short, neither mystery has been explained, and both the nuclear physics and the stellar structure models cannot be blamed for causing the mysteries. One strong candidate for understanding the discrepancy is that the neutrinos have three different "flavors," and on their way out of the sun these flavors change, allowing us to observe only about one-third of the true solar emission.

The solar neutrino flux correlates to solar activity indices with a value of about .27 (a correlation of 1.00 would be perfect). If we repeat the experiment many times, a chance correlation this high would be expected in less than 5% of the experiments. The statistical significance of these experiments is valid only if the experiments were carried out properly. If observed neutrino varia-tions are not solar phenomena, they may be experimental variations arising from the difficulty of making the measurement. If they are real variations in solar neutrino power flux, they are equivalent to about 1 W/m^2 over the solar cycle. This value is less than the 2.1 W/m^2 change in photon irradiance. The combined photon and neutrino flux changes imply a change in total energy flow per unit area per unit time of 3.1 W/m^2 at the Earth's orbit.

Are neutrino flux changes real? Perhaps, but they remain unexplained and only emphasize that even a well-behaved average star like the sun is complicated and, in many respects, still poorly understood.

Variable Stars

In the mid-1970s, P. R. Wilson of the California Institute of Technology asked himself: Can activity cycles like the ones seen in the sun be detected in other stars? If one considers the sun as a star and measures it using the core of the Ca II K line, the solar cycle will be detected. In other stars, the Ca II K line is bright enough to find stellar activity cycles. In succeeding years, several scientists have pursued P. R. Wilson's pioneering line of research. Figure 3.17 provides an example, covering the last 25 years of variations in brightness and

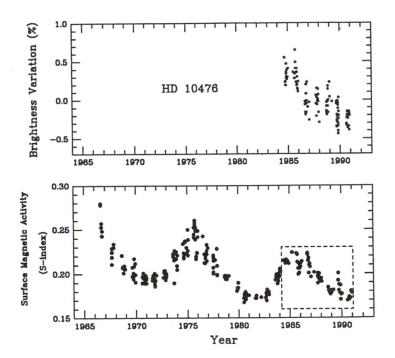

FIGURE 3.17 Brightness and solar magnetic activity variations of the star HD 10476 (from Zhang et al., 1994, with permission). The dashed box in the lower panel shows the time covered in the upper panel. Note the roughly 11-year cycle in this star's magnetic activity. The different levels of magnetic activity at the three activity minima may imply that the star's brightness differs from one minima to the next.

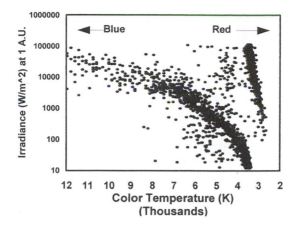

FIGURE 3.18 The Hertzsprung-Russell diagram for nearby stars, with the stellar irradiances (W/m²) plotted as a function of stellar color temperatures (°K). The sun *(open square)* is a main-sequence star with a G2 color classification, a color temperature of 5780 °K, and an irradiance of 1367 W/m² at 1 astronomical unit. A few very hot stars (temperatures greater than 12,000 °K) are not plotted.

solar magnetic activity, for another star (HD 10476). This star is slightly redder than the sun (class G5), more active, and has a greater change in its brightness over its cycle.

In recent years, stellar cycles with periods ranging in length from about 7 to 20 years have been detected in most solar-like stars. Not only are astronomers detecting changes in stellar activity, they are also detecting changes in stellar luminosities. However, first let us examine the types of stars that exist.

Stars differ from each other in both their color and their brightness, or luminosity. Plotting stellar luminosity versus stellar color reveals definite patterns. This type of plot is known as the Hertzsprung-Russell Diagram after its two originators (see Figure 3.18). Our version plots the stellar total irradiance measured at 1 astronomical unit versus the color temperature of the stellar photospheres in degrees Kelvin. Most main-sequence stars fall along a curving line stretching from the upper left corner to the lower right corner. The sun is a main-sequence star with a color temperature of 5880 °K and an irradiance of 1367 W/m². Another large group of stars, known as red giants, follows a straight line along the right side of the figure. In addition to the variations in color temperature, there are color changes, with blue stars on the left and red stars on the right. Star colors are designated with the letters OBAFGKMN. These color classifications range from blue (O) through white (A) to yellow (G) and then to red (M and N). Stars range from very hot (O) to cool (N). The sun is a G-class star, which is often designated as G2 to indicate it is about two-tenths of the way between a G0 and K0 star.

Five billion years ago, the sun began as a late G-class star, perhaps G7 to G9, with an initial irradiance at 1 astronomical unit of about 1000 W/m² and a color temperature of approximately 5400 K. Since then it has steadily warmed up, with a 30% increase in luminosity, and its color has changed from reddish to yellow. In the future, the sun will brighten even more and eventually will move off the main sequence and become a red giant star. These evolutionary changes occur because nuclear fusion reactions use up the hydrogen fuel.

Stars younger than the sun are more active. These red stars often have prominent starspots that last for many years and cover significant portions of their surfaces. As main-sequence stars evolve, their activity gradually decreases and may eventually cease altogether. Some stars become brighter with increased activity, as the sun did in its last two cycles, but other stars become dimmer. For younger stars, sunspot blocking seems to dominate facular emission. Most stars vary at some level. Figure 3.19 shows the results of the Hipparchus satellite experiment that detected variations in light output of numerous stars. The main sequence reveals that stars both brighter and dimmer than the sun display more variability than the sun itself. Figure 3.20 replots the results of Figure 3.19 to show this more clearly. Even though the sun varies, it is one of the more stable stars.

Figure 3.21 reveals yet another way to look at stellar activity and luminosity variations. Here, three stellar variables are plotted against the mean level of

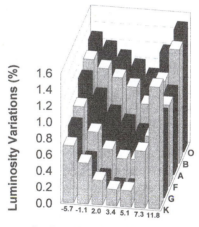

Stellar Absolute Magnitudes

FIGURE 3.19 A Hertzsprung-Russell diagram of stellar brightness variations. From left to right, stars are plotted from the brightest to the dimmest, expressed by their absolute magnitudes. From front to back, the stars are shown in six layers for the KGFABO color classifications, which correspond to stellar colors from red to yellow to blue. The vertical bars show the relative luminosity variability. Note that nearly all stars, including solar-like stars, show some brightness variations. The sun is in the second layer (class G, yellow) with an absolute magnitude of about 5. Compared with most stars, the sun displays low variability. These measurements are derived from the Hipparchus satellite and show the relative variability in stars over several months. (Data from Eyer et al., 1994.)

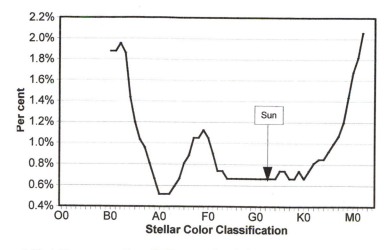

FIGURE 3.20 A Hertzsprung-Russell diagram of variations in brightness of stars along the main sequence expressed as a percentage of its brightness. Although a variable star, the sun fluctuates less than most. As in the previous figure, these variations are shown over several months. For the sun, the largest variation is about 0.5% when a large sunspot passes across the solar disk. (Data from Eyer et al., 1994.)

chromospheric activity. In the upper panel we note that as stellar activity increases, the light output of stars becomes more variable. The sun's level of activity is about average, but its variations in brightness are well below average. This suggests that in the last two solar cycles, we have only seen a small portion of the brightness variations we would see if we observed many solar cycles. The middle panel tells us that the sun's short-term activity variations and cyclic variations are normal for main-sequence stars. The final panel shows that the sun is on the borderline between having its output dominated by sunspot blocking or by facular emission. In the last two cycles, facular emission has dominated sunspot blocking, but the lower panel questions whether this is always the case. Huge sunspots have been detected on the surface of red stars. As these sunspots rotate with the star, the star's light output is modulated in a way that suggests sunspot blocking predominates. During its early life, the sun resembled these stars. At one time, solar activity and solar brightness were anti correlated. As the sun evolved, the present positive correlation replaced this anticorrelation. When this transition occurred is not clear, but is generally assumed to have taken place many millions of years ago.

Solar Constant versus Solar Luminosity Variations

The sun's output varies not only with wavelength but also with heliographic angle. This raises the important distinction between solar constant and solar luminosity. The solar constant is the sun's total irradiance measured in the di-

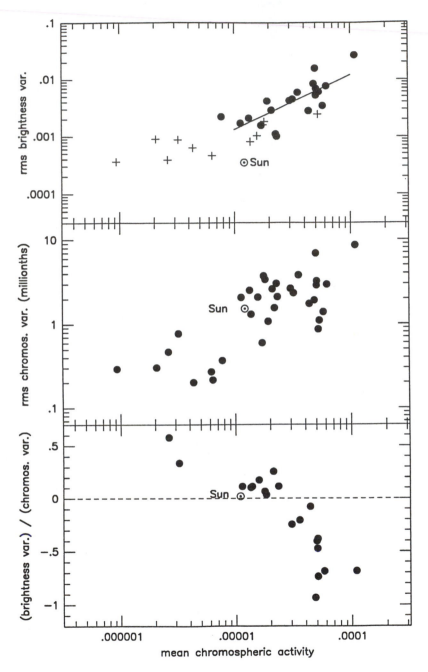

FIGURE 3.21 Variations in stellar brightness, root-mean-square chromospheric variability, and their ratios, all plotted as functions of the mean chromospheric activity. In the lower panel, the sun borders on having its brightness positively correlated with activity for the last two cycles and displaying a possible anticorrelation. (From Lockwood et al., 1992, with permission.)

rection of the Earth corrected to 1 AU. The solar luminosity is the total energy output of the sun, theoretically including all energy forms (neutrinos, solar wind, etc.). Superficially, these two energies might seem to be related simply by a factor equal to the area of a sphere with radius 1 AU. This is not true because the sun is not spherically symmetric, producing extra solar output at certain heliographic angles (polar latitudes) compared with other angles (equatorial regions).

Both Lockwood and Skiff (1990) and Baliunas and Vaughan (1985) found that the variations in the total radiative output from solar-type stars exceeded the currently observed solar-constant variations (from spacecraft over the last decade) by a factor of nearly 4. Excluding remote alternatives, this suggested the following:

1. The sun may undergo irradiance variations several times larger than any we have seen during the past decade.
2. Compared with other solar-analog stars our sun is highly unusual because it has especially quiescent radiative output.
3. Our terrestrial position in the heliosphere (the Earth always lies close to the sun's equator, since the tilt of the sun's rotation axis, B0, relative to the ecliptic plane is a small angle, 7.25°) provides a special vantage point that reduces the observed solar-irradiance variations.

Examining this third possibility, Schatten in 1993 attempted to explain why the sun's solar-constant variations appear especially docile compared with other stars. He found that the fourfold irradiance enhancements seen in other solar-type stars compared with the sun could be explained by the special viewing angle from which the Earth views the sun (see Figure 3.22). Since active regions tend to be at low heliographic angles (HA) and sunspots are predominantly dark near the disk center, Earth sees sunspots preferentially (top portion of the figure). When viewed at high HA, the faculae (bright solar features) are seen better (lower portion of the figure). At low HA, our sun happens to have a close balance between the bright emission from faculae and the dark, reduced emission from spots. Higher HA show significantly more variation in activity (up to a factor of 6). Overall, due to the extra emission seen at high HA, which is not sampled because of the Earth's preferential low latitude at which few solar-constant variations are seen, the solar luminosity could vary about three times as much as the solar constant.

Cyclic Variations in Solar Brightness: The Implications for Weather and Climate

Some highlights of this chapter are:

- Total solar radiative output varies in an 11-year cycle.
- The variations are stronger in the ultraviolet and blue spectra than in the red and infrared spectra.

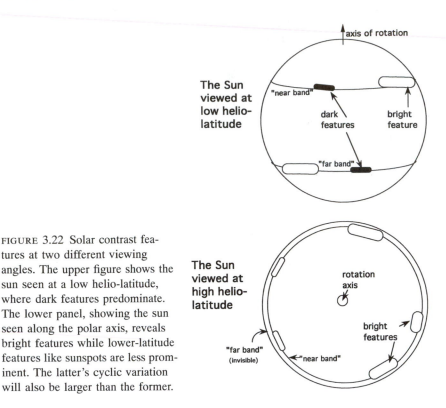

FIGURE 3.22 Solar contrast features at two different viewing angles. The upper figure shows the sun seen at a low helio-latitude, where dark features predominate. The lower panel, showing the sun seen along the polar axis, reveals bright features while lower-latitude features like sunspots are less prominent. The latter's cyclic variation will also be larger than the former.

- In recent years, extra facular and active-network emissions have dominated over sunspot blocking deficits.
- Regression models for solar variations can be constructed going back many years.
- Both the solar photon and neutrino energy fluxes vary on the time scale of the solar cycle, although the latter is questionable.
- Most stars show some variation in their light output.
- The sun may be atypical among stars because it displays less variability than similar stars; this may be due to the special angle from which we view the sun.
- Some failures of the regression models, variations in neutrino output, and variations in line widths may indicate that other, unaccounted-for, phenomena are taking place.

In 1910 W. J. Humphreys wrote:

In addition to a careful determination of the solar constant and terrestrial temperatures during one or more spot cycles, it would be well to measure, at the same time, the accompanying changes in the ultraviolet portion of the radiation, and also to follow, over the same cycles, the temperature and height of the isothermal layer, and to note, if possible, the amount of ozone in the upper atmosphere. The information here called for is difficult, though not impossible to obtain; but much of it—it may be all—is essential, though perhaps not sufficient, to the solution of the complex problem concerning the relation of solar activities to terrestrial temperatures—a problem of great interest, both from the strictly scientific and from the purely utilitarian standpoints.

Scientists have met some, but not all, of Humphreys' urgings. The variations in solar irradiance seem to follow an 11-year cycle. For now, we assume that the sun's output varies, much like the Foukal and Lean model. What effect will these variations have on climate? We devote the next six chapters to this question. Since we have established that the sun is variable star, let us examine variations in climate. We look first at exactly what "climate" means and, given an 11-year solar variability, what we can hope to find in the climate record. Several chapters will also be devoted to trying to find out if empirical climate studies make sense.

II. THE CLIMATE

4. *Climate Measurement and Modeling*

Having considered the sun and its variations, we now turn to Earth's climate and climatic variations. We examine the definition of climate and the difficulties in measuring it. Awareness of these complexities is critical for an appreciation of how difficult it is to demonstrate changing climate. Separating trends from random variations is the first step in defining climate change.

After reviewing the statistical properties of climate, we deal with theoretical climate models. This background is important for understanding how solar variations might affect climate. The following four chapters review specific sun/climate relationships, and the statistical and physical guidelines developed now will be used to select pertinent studies.

As the heat source that drives Earth's climate, the variable sun is important when studying climate change. With many, if not most, modern popular accounts focusing on how humanity is altering climate, it is important to realize that solar variations may play a significant role in the background natural variability. To understand anthropogenic (human-made) influences on climate change, we must be able to make distinctions among the contributions that arise from naturally occurring climate variability. Natural climate variations include a possible solar-irradiance component.

Man-made climatic changes are not well known, and natural climate variations are uncertain too. For example, we do not know whether a man-made doubling of atmospheric carbon dioxide provides a 1.5 or a 4.5 °C increase in mean global temperature. This uncertainty arises, in part, because natural climate variability acts as "noise" to confuse our measures of man-made influences. To obtain accurate results, we must understand and remove these background noise sources. Although these temperature changes seem small, they can have tremendous global impact on the survivability of species and on many different aspects of life. In addition, the uncertainty factor of 3 is highly important because it tells us that the risk in emitting a quantity of carbon dioxide is uncertain by this same factor.

The natural climate variability arises from several sources:

1. solar variations
2. albedo changes (due to the changing landscape, e.g., deforestation, ice cover, etc.)
3. the inherent variability of the chaotic climate system (year-to-year fluctuations in cloud cover, etc.)
4. volcanic influences
5. other effects

The sun's role in past climate change is not well understood. Some researchers believe the sun plays the major role in natural climate variations; others say it's a negligible role. We favor the viewpoint that solar activity has been increasing in this century in a way that appears to fit the global temperature record, a fit that is perhaps better than the variations in carbon dioxide. The negative view insists there are no significant solar influences because solar variations are too small. Nevertheless, the sun can certainly be expected to have had a more prominent role in past climate change than man-made influences because mankind is a relative newcomer to the Earth. Similarly, the growing evidence for the sun's control of past climate should not be understood to suggest that anthropogenic changes are now unimportant. Rather, to understand climate change we need to study both greenhouse gases and solar influences. The climate system is more complicated than a direct one-to-one relationship between carbon dioxide and temperature. We will return to these other theories of climatic change in chapter 11.

What Is Climate?

Most people consider climate merely the mean or average weather. Others, being more specific, may say it is the average temperature or average rainfall. These definitions are somewhat vague. First, consider a specific location. With sufficient resources one can measure many meteorological variables. Let us consider one example: air temperature at specified times each day. Alternatively, one could measure the maximum and minimum daily temperatures. Already we would have at least four variables that could be used to define climate: (1) the daily mean temperature, (2) the daily maximum temperature, (3) the daily minimum temperature, and (4) the daily temperature range. Measurements taken over 30 years could create a mean for each parameter and a climatology at this location based on these same variables. A 30-year standard period is usually adopted for these means. Every 10 years a new 30-year base period is created by shifting the original base forward 10 years. Measurements differing from the base mean are called anomalies.

The Earth's surface temperature is only one climate variable. Barometric pressure, wind velocity, wind direction, and relative humidity are other common climate parameters. Information on atmospheric composition such as the total column and surface ozone, the total column of water, and the carbon dioxide concentration are also climate variables. The aerosol loading, size distribution, and absorption properties are atmospheric composition parameters. Vertical profiles of all these quantities are climate parameters. Climatologies of rainfall, hail, snowfall, and occurrences of lightning or thunder are useful, as well. A cloud climatology involves measurements of total cloud cover, as well as cloud types, heights, thicknesses, temperatures, and pressures. The duration and the percentage of possible sunshine are related quantities. Other parameters akin to radiation include visibility, the sun's normal incidence radiation, the sky radiation and its polarization, the global radiation, the reflected solar radiation, the thermal radiation to and from Earth's surface, and the spectral distribution of all these radiation components.

Surface parameters include albedo, snow cover, soil moisture content, the Palmer drought index, and soil temperature. Beyond these single-site measurements are droughts, and their severity, river flows, lake levels, rate of entropy production, and so on. A complete list of possible variables requires several pages. In general, we must consider the temporal and spatial sampling of energy, matter, information, and boundary conditions. These numerous variables are mentioned simply to show how complicated climatology becomes when one considers all possible parameters. Temperature, pressure, wind velocity and direction, precipitation, and relative humidity are probably the most commonly measured climatology variables. Thirty-year means for these quantities are usually referred to when discussing climatology and climatic change.

Difficulties in Measuring Climate

Considerable difficulty occurs when measuring even the most common climate quantities, and even greater difficulty arises when comparing different stations at different times. Part of this problem occurs because measurement procedures differ from country to country and evolve slowly as new measurement techniques develop. These problems need to be considered because climatologists must try to decide whether a climate change is real or only an error caused by inconsistent measurements.

To simplify matters, we consider only the problems in constructing a temperature climatology. Temperature is the most commonly, and presumably the most accurately, measured parameter. If we encounter trouble measuring temperatures, other variables surely present even greater difficulties.

In 1953, J. M. Mitchell Jr. divided measured temperature changes into two categories, apparent and real. An apparent change arises through some fault in the measurement process. Examples of apparent changes are:

- The method of computing daily means could change: for example, using 24-hourly values rather than the average of the daily maximum and daily minimum. There may be systematic differences between these two methods and other methods that, if not corrected, could be misinterpreted as a climatic change.

- The thermometer used could change due to replacement for breakage or aging. This problem is less likely now than 100 years ago.
- The thermometer shelter could change, either through design or condition. Identical thermometers placed in different shelters next to each other can give different readings. Replacing an entire network of shelters might produce an apparent regional climatic shift, although none actually occurred. The shelters need not even be replaced to produce such erroneous events. Shelters are painted white and if the brand of paint is changed for the entire network from an oil-based to a latex pigment, the paint's emissive properties might cause a shift in the mean temperatures. The effect would be the same as a change in shelter structure, but would be more difficult to track. Careful records are made of changes in shelter design, but no records may be kept of changes in the kind of paint used.
- Even small changes in thermometer location and surroundings can cause an apparent change in climate. About 1930, in Denver, the thermometer was moved from the first to the second floor. During the following years, the recorded temperatures decreased slightly and cold records increased significantly. This change might easily be missed by analysts seeking climate changes. Another change might occur simply because a nearby tree or group of trees subtly altered the local heat balance and downwind temperature. Again, the local change might be interpreted as part of a regional change.
- Mitchell lists some changes as real, but these changes are still caused by local effects, not global effects. Foremost among real changes is the urban heat island effect. This effect is usually attributed solely to fuel combustion whose waste heat causes urban heating. As early as 1850 the frontier city of St. Louis had an urban heat island that occurred not from fuel combustion but from the exterior surfaces of buildings so constructed that they acted as light traps. Sunlight that is normally scattered back to space is instead multiply reflected from the buildings until it is absorbed. Many nearby buildings can act as absorbing cavities which lead to urban heat islands. Because of this effect, even small towns with no industry can be warmer than the surrounding country. Separating these spurious effects from other real climatic changes is very difficult. The urban heat island has contributed an estimated 0.1 °C increase in the average hemispheric warming of about 0.5 °C observed during the last century, creating a spurious warming trend.

Along with the difficulties in constructing a homogeneous temperature record at one site are complications that arise when averaging several stations together. Since stations are not uniformly spaced, they must be area weighted. Different weighting techniques can produce different reconstructions of regional temperature changes. The largest spatial scales create the most difficult problems when regions such as the oceans are not sampled or are poorly sampled. Small measurement errors at a location representing a large area could adversely influence the values of hemispheric or global trends. In 1992 Gunst and his colleagues looked at several different techniques of area-weighing temperatures to derive the mean hemispheric variations. Figure 4.1 illustrates the differences between two techniques. The trend here ranges from about 0.05 to 0.15 °C, which equals 10% to 30% of the claimed warming in the last hundred years—a significant effect. Thus, spatial averaging can potentially introduce an artificial warming signal.

Despite these difficulties, several groups of climatologists have reconstructed temperature variations from about 1850 to the present. Reconstructions before 1850 become increasingly difficult, as there are fewer and fewer measurements. Figure 4.2 shows four reconstructions of the Northern Hemisphere temperature anomalies. Although all reconstructions start with the same basic station data, their final results differ slightly. Differences arise from how the individual station data are adjusted for local changes and temperature sampling changes, from the different criteria used to eliminate different stations, from different methods of area weighing, and so forth. The four reconstructions have the same long-term variations but differ in detail from year to year. Even the agreement of the secular variations does not guarantee that the temperature

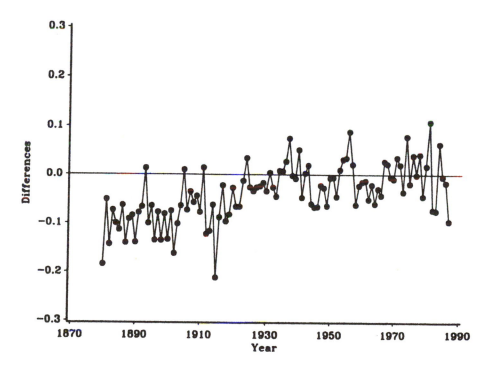

FIGURE 4.1 The differences in temperatures for the Northern Hemisphere using two different spatial weighing techniques. Improper spatial averaging may introduce a false upward trend in temperature. (From Gunst et al., 1992, with permission.)

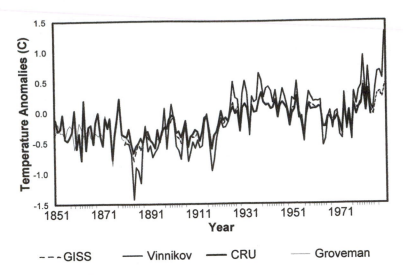

FIGURE 4.2 Four different reconstructions of Northern Hemisphere temperature anomalies from 1850 to the present. Year-to-year details differ but the secular trends agree. Discrepancies arise from slightly different input data and from the various analysis techniques used.

reconstruction is correct because many steps in the analysis techniques are the same and the reconstructions are not truly independent.

When considering climate change, we must be aware of the limitations of our knowledge. Both spatial and temporal gaps exist in our measurements. The measurements themselves may not necessarily be internally self-consistent. Regional, hemispheric, and global representations of climate change may not be accurate. Data limitations can obscure the theories we are seeking to prove or disprove. Solutions are being sought for all these difficulties, so every few years new reconstructions of past climatic changes are published. Recently, more refined temperature records revealed the measured temperature decrease from the mid-1930s to mid-1960s, as shown in Figure 4.2, has become less and less prominent. It is likely that climate reconstructions are varying faster than the climate itself.

Let us now consider some of the practical, significant temperature changes shown in Figure 4.2. Over a century, the mean hemispheric changes equal about 0.6 °C, equivalent to a downward alteration in altitude of about 300 feet or a southward move of about 60 miles. These changes affect agriculture and the optimum location of wildlife habitats, but would be difficult to discern for the typical urban inhabitant. These same changes are equivalent to moves of less than a mile per year north and south. In contrast, many people move many hundreds of miles an average of once every 5 years. The climate changes are important to societies and large collections of people who effectively remain fixed, but are considerably less important to the mobile individual.

Variations about the Mean

The mean value of temperature (or any other meteorological variable) is not a full description of climate. The variability about the mean is also important. For example, two locations with identical temperature means, but one with a large annual amplitude in temperature variability and the other with little annual variations, will be recognized as different, not only by people but also by plants and animals. Locations near large bodies of water, such as ocean islands, often display minor temperature fluctuations compared with mid-continental locations like North Dakota. This variability can be expressed in several ways, such as the extreme differences, mean differences, or standard deviations.

Let's consider temperature variations expressed in terms of standard deviations for three spatial scales: a local mid-latitude site, a continental-size region, and the globe. On a local scale, even the passage of storm systems can cause considerable day-to-day variations, and traveling a short distance can lead to a location with different weather. Averaging spatially, the variability becomes smaller and smaller, as illustrated in Figure 4.3.

Averaging over longer intervals also leads to less variability. Local sites will display yearly averages more similar to each other than will any monthly or single-day average. Longer-period (such as yearly) means smooth the shorter time-scale fluctuations. Local daily variations in temperature are roughly 40 to 50 times greater than the global yearly changes. What does this have to do with climate change? If the sun or any other external source induces climate change, this forcing will typically be a few tenths of a degree, probably relatively inde-

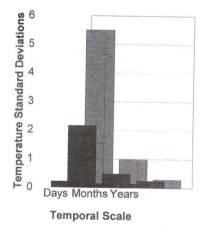

FIGURE 4.3 Typical variations in degrees Celsius for local, regional, and global spatial scales and for daily, monthly, and yearly averages *(back to front)*. A mid-latitude site is chosen for local regions. The global daily and monthly variations are estimated from the Microwave Sounding Unit (MSU) measurements. Compared with the global yearly variations, variations are much larger for short temporal averages and small spatial regions. External forcing of climate will be detected most easily at the larger spatial scales with averages of a year and longer.

pendent of the temporal and spatial scale. In practice, if the sun (earth's axis tilt) controlled 100% of the global yearly variations, only about 2% of the mid-latitude local daily variations could be attributed to this external forcing. In fact, probably no more than about 40% to 70% of the global yearly temperature fluctuations could ever be attributed to any single causal mechanism. From 30% to 60% of these variations are randomly chaotic and unpredictable. Therefore, at a mid-latitude location only 1–2% of the daily variations could be caused by any external forcing. Some sites may have fluctuations parallel to the global variations, while other sites have no variations paralleling the large-scale climatic changes. Many meteorological parameters, such as precipitation, have much greater spatial and temporal variability than does temperature. These variables require longer measurement periods to detect secular trends.

These points are important when considering sun/climate relationships. While many empirical studies consider daily fluctuations and single sites, these deserve less serious consideration than broader studies. The noise level at single sites is so large, it can obscure any solar forcing signal. Hundreds of years of observations would be required at a single site to discover data comparable to a few decades of global or hemispheric mean observations. In sun/climate relationships, empirical studies with any chance of success will tend to consider variations on the largest spatial and longest temporal scales possible. These considerations will also be examined here.

Climate and Weather Extremes

Although changes in the mean values are obscured by chaotic variations, examining the frequency of extreme weather events can be a powerful technique for detecting secular climate variations. For an unchanging mean temperature, a Gaussian distribution can approximate the fluctuations about the mean (see Figure 4.4). If a sudden small change in climate equal to 0.1 standard deviation occurs, the entire Gaussian distribution shifts. Considerable time will be required for the new climate condition to fill in the new Gaussian curve. Even after fully sampling both climates and producing two Gaussian curves, it is difficult to tell if the change has occurred. If we focus on the extreme tails of the two Gaussian curves, however, we can see large changes in the ratios of the two curves. Table 4.1 summarizes the changes in the frequency of events 3 standard deviations from the mean.

Small changes in the mean climate amplify the probability of seeing extreme events more frequently. This effect is interesting for several reasons. It is often easier to count the number of extreme events than it is to measure the small shifts in the extremes. On the other hand, small changes in site locations can lead to large changes in the number of recorded extreme events. In the Denver example mentioned earlier, moving the thermometer upward by 10 feet (3 meters) to the second floor produced more common record lows. Changes in the number of extreme events do not necessarily indicate a climate change, but they do suggest a possible change at the site. Because extreme events are infrequent, long intervals are needed to detect changes in their variability. Extreme

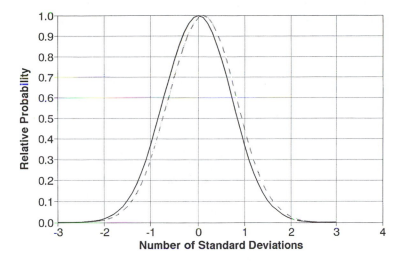

FIGURE 4.4 A plot of two Gaussian curves that are shifted relative to one another by 1 SD. At half maximum, the full width of a Gaussian curve is 1.665.

events have no statistical advantage over seeking changes in the mean values. The most important aspect of changes in the frequency of extreme events may reside in the economic sphere. Most adverse climate effects reveal themselves through extreme changes. Favorable climate changes may dramatically diminish the number of adverse extreme events, so climate change may be either adverse or beneficial, depending on location. Nevertheless, since many species are adapted to preexisting conditions, changes are more likely to be unfavorable to them.

What is the probability of an extreme event? In the absence of any climate change, during any one year the probability that one will experience the warmest day ever seen in N years is simply $1/N$. If one measures for 100 years and the climate remains stable, each year on average will have 3.65 extreme warm days and 3.65 extreme cold days, with the extremes evenly distributed over

TABLE 4.1 Changes in the frequency of the events 3 standard deviations from the original mean when the mean shifts by 0.0 to 0.5 standard deviation.

Shift	Frequency
0.00	1.00
0.10	1.80
0.20	3.19
0.30	5.53
0.40	9.39
0.50	15.64

time with no change in climate. Therefore, secular changes in climate could manifest themselves as secular changes in the frequency of extremes.

Consider, for example, the temperature record at West Lafayette, Indiana. Figure 4.5 shows the number of extremes from 1901 to 1980. Without any climatic change, 4.56 extremes of each type are expected each year. The number of extremes fluctuates from zero to more than 20. Such wide-ranging values are expected because in this case the standard deviation of the probability of extremes is 8.95 events per year. Theoretically, we expect about 4 years out of 80 to have 20 or more extreme days. At West Lafayette, only the year 1936 deviates so far from the average, suggesting that temperatures here are actually a little less variable than predicted by stochastic theory. In the absence of climatic change, the expected number of records from 1919 to 1946 is 126. There occurred a record maxima of 175 and a record minima of 68, suggesting a warmer than normal period. Figure 4.2 reveals that the Northern Hemisphere was warmer, and West Lafayette's trend in extremes follows the hemispheric variations. Therefore, during sun/climate studies, there may be some value in looking for changes in the frequency of extremes that correlate with changes in solar activity.

Statistics in Sun/Climate Investigations

We have examined the means and extremes of climate, as well as some difficulties in determining past climatic changes. We have argued that finding a sun/climate relationship usually has a better chance of success when using large quantities of spatial and temporal data. Empirical studies of sun/climate relationships require a knowledge of statistics. Several guidelines can increase the

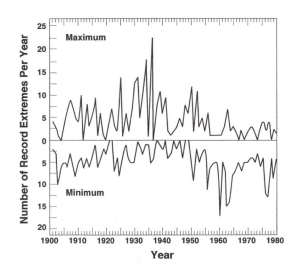

FIGURE 4.5 The number of daily record maximum and minimum temperatures at West Lafayette, Indiana, from 1901 to 1980. (From Agee, 1982, with permission.)

probability of success in these studies, such as (1) pursuing studies in which a physical basis exists, (2) understanding the limitations of both the terrestrial and the solar data time series, (3) carefully examining the significance of the results and not overestimating significance by failing to consider persistence or common periodicities in the data, and (4) considering the practical importance of the findings by objectively accessing the fraction of the variations that are solar induced. Here we hope to recognize and follow these guidelines when selecting material. To follow these guidelines requires a knowledge of both statistical problems and the physical basis of climate. Appendix 3 presents more details concerning statistical methods, including a brief technical discussion of some statistical techniques and problems that have arisen in sun/climate/weather investigations. Bayesian estimation, Chree analyses, and cross-spectral analyses are commonly used and useful tools. Also mentioned are some statistical problems leading to false conclusions by both the "skeptics" and the "believers" of sun/weather/climate research. Skeptics may falsely conclude that the "disappearance" of an effect eliminates its possibility. Believers, on the other hand, may overstate their positive claims through subtle, inappropriate uses of statistics. As skeptics and believers may reach different conclusions concerning the significance of statistical study results, care must be taken because the mathematical basis of statistics assumes independent measurements. If spatial or temporal correlations enter the data, these same data are not independent and the study's significance must be reduced accordingly. Having discussed some aspects of the statistical basis of climate, we now turn to its physical basis.

Earth's Radiation and Energy-Budgets—Basic Principles

This section briefly reviews the consequences of solar radiative changes on the Earth's radiation and energy budgets, followed by a more comprehensive discussion of climate theory and modeling.

With no climatic change, the Earth's energy input equals the reradiated energy. Most energy inputs are from the visible or shortwave radiation. The reradiated or emitted energy is called the thermal or longwave radiation. More than 99% of the shortwave radiation has a wavelength less than 4 microns with a peak at 0.5 micron (see Figure 3.3). Above 4 microns nearly all the longwave radiation is in the infrared, peaking at about 15 microns. Figure 4.6 shows the shape of the thermal spectrum.

Clouds, the atmosphere, and the Earth's surface reflect about 30% of the incoming solar radiation back into space. Therefore, the Earth's albedo is 0.30. The remaining 70% of the radiation is absorbed or thermalized to heat the Earth's surface; it evaporates water (latent heat), or causes convection (sensible heat). The absorbed energy is eventually reemitted as longwave (or infrared) radiation, with 6% coming from the surface and 64% from the clouds and atmosphere. This whole process is known as the Earth's heat engine. The calculation or measurement of the various paths for the radiation and energy flows are the Earth's radiation and energy budgets, respectively.

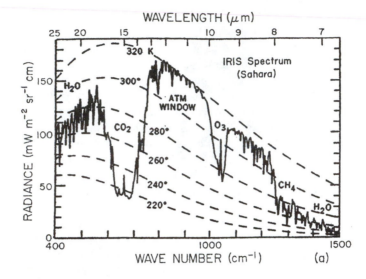

FIGURE 4.6 A typical thermal radiation spectrum that would be seen by looking at the Earth from space. (From Hansen et al., 1993, with permission). Note that the wavelength scale at the top of the figure is in reverse order. Strong absorption bands at 15 microns due to carbon dioxide and at 9 microns due to ozone cause the Earth to warm. Water vapor is the most important greenhouse gas, trapping thermal radiation about a hundred times more effectively than carbon dioxide.

Figure 4.7 shows the Earth's radiation budget with the incoming solar irradiance in yellow and outgoing thermal radiation in red. In chapter 3 we saw that, for the last solar cycle, the solar irradiance was about 1368 W/m² when the sun is quiet and about 1370 W/m² when the sun is active. In energy budget calculations, the solar irradiance is divided by 4 to account for the fact that the Earth radiates as a sphere but only absorbs a "circle" of solar radiation. The surface area of a sphere is $4\pi r^2$, whereas a circle has an area of πr^2. Dividing 1368 by 4 yields 342 W/m². Thus, the Earth absorbs and emits 239.32 W/m² per unit area, on average, when the sun is quiet (taking into account the albedo). When the sun is active, this number is 239.75 W/m², or 0.43 W/m² greater. A blackbody is defined as an object in thermal equilibrium. It emits radiation in the most efficient manner possible, given its temperature. The intensity of the blackbody radiation varies with the fourth power of its temperature ($I = \sigma T^4$). Because of the fourth-power law, a fractional change in temperature T causes a fractional I that is four times the fractional change in T. A greater emission of radiation implies a greater temperature, and, treating the Earth as a blackbody, its new effective temperature would be 255.11 °K or 0.11 °K warmer when the sun is active. Most climate radiation models reveal the same sensitivity of the Earth's temperature to changes in the solar constant. To a first approximation, the blackbody spectrum is similar in shape to the radiation emitted by the Earth.

The situation is actually more complicated than the one presented here.

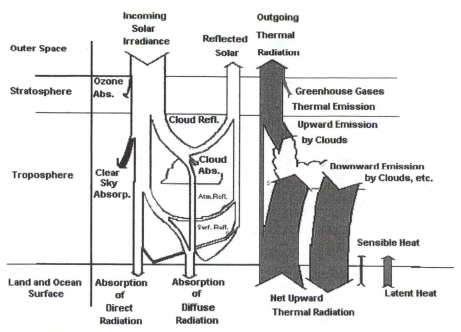

FIGURE 4.7 The Earth's radiation budget showing the major paths of the incoming solar radiation *(left, shaded)* and for the outgoing thermal radiation *(right, black)*. For a quiet sun, 342 W/m^2 is input and about 102.7 W/m^2 is reflected. To achieve radiation balance, 239.32 W/m^2 is absorbed and reemitted. Under these conditions, Earth's equilibrium temperature is 255 °K. Thirty percent of the incoming radiation is reflected back into space by the atmosphere and surface, so the Earth's albedo is 30%. The table below lists some approximate values (in Watts per square meter) for the radiation budget for the quiet sun (QS) and active sun (AS) and their differences (Diff).

	QS	AS	Diff.
Incoming solar irradiance			
Absorbed by ozone	10.3	10.4	0.1
Clear sky absorption (mostly H$_2$O)	58.1	58.2	0.1
Direct radiation absorbed by surface	75.2	75.4	0.2
Diffuse radiation absorbed by surface	71.8	71.9	0.1
Total solar radiation absorbed by surface	147.0	147.3	0.3
Cloud absorption	23.9	24.0	0.1
Cloud reflection	65.0	65.1	0.1
Total reflected solar radiation	102.7	102.9	0.2
Outgoing thermal radiation			
Upward emission by clouds	219.9	220.2	0.3
Greenhouse gases thermal emission	10.3	10.4	0.1
Downward emission to surface from clouds, etc.	328.3	328.8	0.5
Upward emission from surface	389.9	390.5	0.6
Net upward emission by surface	61.6	61.7	0.1
Sensible heat (convection)	17.0	17.0	0.0
Latent heat (evaporation)	82.1	82.2	0.1
Total outgoing thermal radiation	239.3	239.6	0.3
Total reflected solar and outgoing thermal radiation	342.0	342.5	0.5
Albedo of Earth	30%	30%	0.0%
Temperature of Earth (°K)	255.0	255.1	0.1

Note that the differences between the active and quiet sun are small and therefore difficult to detect experimentally. Since sunlight is bluer when the sun is active, the Earth's albedo can fall below 30% due to strong ozone absorption in the ultraviolet.

First, these calculations assume the Earth has reached radiative equilibrium for both the quiet and active sun. If the Earth has a thermal-response time constant of perhaps a few years with an imposed 11-year solar forcing cycle, it would not reach full equilibrium in 11 years. Nonetheless, it would be close to equilibrium, requiring reduction of the 0.11 °K value to perhaps 0.1 °K. Second, when the sun is active, the Earth's albedo may be lower than when the sun is quiet. This effect could arise because the active sun produces more ultraviolet radiation and the Earth is darker at these wavelengths. Consequently, more radiation than 239.75 W/m^2 will be absorbed. This number could rise to about 240 W/m^2. In this scenario, Earth's cyclic warming would be 0.18 °K. Third, there may be several feedback effects that increase the warming, including increased water vapor that traps thermal radiation and causes a warming through the greenhouse effect. Figure 4.7 suggests that we can expect an additional 1% increase in latent heat or a 1% increase in both evaporation and precipitation. If the feedbacks are similar to those predicted for the carbon dioxide greenhouse effect, the 11-year cyclic temperature variations could be about three times stronger, or 0.33 to 0.54 °K. If the entire atmosphere warms by this amount, the Earth's surface temperatures would increase accordingly. These higher numbers are close to recent satellite measurements. This is a subject to which we will return in the next chapter. A fourth effect that could magnify solar influences on climate, and is related to the way solar ultraviolet variations affect the middle atmosphere region called the stratosphere, is covered in the next three subsections.

Climate Modeling: The Nature of the Problem

This section covers the nature of the climate "problem." The next section deals with the general problem of solar influences. In a later section dealing with climate modeling, we discuss how solar UV variations may play a significant, if overlooked, role in climate change.

To ascertain a planet's climate, it is necessary to obtain the long-term average or steady-state conditions. This assumes there is no variation in mean conditions. In particular, one is interested in solving the energy balance equation:

$$\frac{\delta E}{\delta t} = 0 \tag{5}$$

where E is the mean energy of the Earth system, a primary driver of the planet's mean temperature. After assuming such boundary conditions as the mean solar constant, albedo, and so on, one attempts to develop a methodology to find an atmospheric temperature structure consistent with the external boundary conditions and to arrive at a solution.

The climate problem can be viewed as an "inverse" problem, although by simplifying assumptions it can be converted into a boundary value problem. Let us first consider the inverse aspects. A "direct" or "forward" problem is one specifying a given set of conditions, such as the thermodynamic state variables of an atmosphere, and attempting to calculate some behavior such as the atmosphere's infrared emission. An inverse problem is the opposite of a direct problem in which an atmosphere's infrared output (e.g., for energy balance equal to the shortwave energy input) is given and the problem is to ascertain the thermodynamic state variables of the atmosphere. Broadly speaking, solving inverse problems involves determining the internal structure or past state of a system from indirect measurements. Such problems include determination of diffusivities, conductivities, densities, sources, geometries of scatterers and absorbers, and prior temperature distributions. "Ill-posed" inverse problems exist in many physical and geophysical areas from tomography to electrical conductivity, scattering, heat conduction, optics, and even the interpretation of atmospheric measurements (see Engle and Groetsch, 1987: McLaughlin, 1984). As in climate modeling, the field of inverse problems is in its early stages, with literature being published in proceedings rather than conventional textbooks. Inverse problems that lack unique solutions (often called "ill-posed") abound and arouse considerable interest because problems with unique solutions seldom create controversy. Different classes of inverse problems exist, and if a particular problem is ill-posed, the specific solution depends on how the problem is solved and what assumptions are made concerning key aspects of the problem.

Even for ill-posed problems, one attempts to obtain a single solution derived from the physics. Given a set of measurements, one sometimes tries to discover the entire class of possible solutions. For example, the uniqueness of stellar envelopes, which we believe are similar to the construction of a climate atmospheric model, was considered by Kippenhahn and Weigert (1990), who note that "multiple solutions for the same parameters . . . exist."

Many inverse problems can be modeled abstractly as $Kx = g$, in which K is a given operator, g is given measurements or observations, and x is the desired solution representing internal parameters or past history, inaccessible to direct measurement. In many cases, the form of K makes the problem ill-posed, engendering peculiar difficulties in the numerical approximations and the interpretation of solutions.

Many examples can be chosen which demonstrate greater complexity. However, let us use a very simple example to illustrate the inverse problem. Given a planet's external gravity field, what is the internal density structure? From the external field, one cannot distinguish a point mass at the center of the planet from a spherical shell of mass anywhere inside the planet, or from many other possibilities. "Solving" the inverse problem typically requires balancing not only the external boundary conditions but also obtaining a particular solution or set of solutions. This necessitates selecting some reasonable physical assumptions because the particular solution is only as good as the accuracy of the underlying assumptions. If a uniform internal density structure for the planet is assumed, a particular solution emerges which usually will be a better approximation than one based on the presumption that the entire mass is condensed in the center. Specific problems do have specific solutions (a planet's density does have a certain form, and Earth's mean climate does follow a specific behavior

pattern). However, the particular solutions may depend on some poorly specified parameters (such as whether the density is uniform) or may be unobtain-
able. A "correct" model reproduces the phenomena it is supposed to describe.
The model must not predict nonexistent phenomena. Generally, the simplest
model meeting these two last criteria is adopted.

The climate problem is not clear-cut, as disagreement exists regarding the
actual boundary conditions, important parameters, key elements, and method of
attack. In particular, we are concerned with the form of the radiation laws. As
with any problem, selecting different laws results in different solutions. However, the solution to the inverse problem might or might not be unique, depending on the radiation laws and other elements. To solve even an ill-posed
problem, one must make various assumptions and define a method for obtaining
a particular solution, known as "removing the nonuniqueness." Here, the form
of the differential equations and method of solution seem to convert the climate
problem into a boundary value problem in which the method of solution and
choice of boundary conditions play a key role in obtaining a particular solution.
This will become evident in the next section, in which we supply two different
solutions based on different variables. The origin of these differences may be
the inherent inverse nature of the climate problem, but the outward manifestations are different choices in radiation laws and their methods of solution.

Solar Influences on Climate

The influence of solar activity on climate is a side branch of the general field
of climate modeling. Many climate modelers feel that with oceans, clouds, meteorology, changing albedos, volcanic effects, Milankovitch effects, and so on,
adding solar activity would only complicate an already overly complex field
and simply increase bewilderment. This confusion appears to arise for two reasons. First, the methodology of climate modeling is not yet rooted in any clear
mathematical foundation. Techniques enable "solutions" to be obtained, but
their uniqueness is not proved. Second, with only fleeting glimpses of one planet's climate, there is little room for comparisons, generalizations, or much theory "testing." Although climatologists have amassed considerable data about
our past climate, the breadth of the data is currently limited to our own planet.
Far less is known about planetary climatology because only a handful of objects
within our solar system are available for study, and all are roughly the same
age and bathed by the same sunlight.

Considerable concern exists that our present-day climate is changing because of the influence of anthropogenic greenhouse gases introduced into our
atmosphere. Some other theories for long-term climate changes include the Milankovitch hypothesis that relates changes in the Earth's orbital parameters to
climate change and which explains the Pleistocene glacial cycles of the last
million years fairly well, volcanic eruptions, man-made aerosols, land surface
changes, and so forth. One recent, controversial theory is that the Earth's
climate has been altered, at least in part, by solar activity variations. Friis-
Christensen and Lassen in 1991 found dramatic support for this option by

showing that proxy temperature records of the Earth's atmosphere closely follow sunspot cycle duration. Less dramatic, but of longer duration and seemingly more significant, are comparisons of Earth's climate history with Eddy's
and Damon and White's 1977 proxy solar activity records. Here, weak but
persistent correlations exist between climatological data and solar activity, as
deduced from the solar modulated isotopes ^{14}C, ^{18}O, and ^{10}Be. Occasional dramatic events distinguish these correlations. For example, the Vikings flourished
in Greenland during the eleventh and twelfth centuries, known as the Medieval
Optimum period, when solar activity was high. The seventeenth century produced a global cooling known as "the Little Ice Age" coincident with the
Maunder Minimum in solar activity. These long-term solar activity influences
may be important contributors to natural climate variability.

In opposition to the above view is the great concern that, based on energy
considerations, known solar variations appear too small to cause the observed
climate fluctuations. Accepting solar influences on climate change requires an
equality between the global temperature variations that have accompanied climate changes and the estimated influence from solar variations. The solar constant is known to vary only about 0.1%, whereas Crowley and Howard (1990)
stated that "inferred climate fluctuations of 1.0–1.5 °C . . . would require solar
constant variations of approximately 0.5–1.0%." On the other hand in 1990,
Wigley and Raper reported that a change of ΔS, in W/m^2, in solar radiation
perturbs the global mean temperature by 0.3 to 1 °K per W/m^2 (derived from
$\Delta T = \lambda^{-1} 0.7/4$ °K, where λ^{-1} is the climate sensitivity). In 1984 Hansen and
his colleagues found that a doubling of CO_2 is expected to warm the climate
by about 2 to 6 °K and that this doubling is equivalent to a 2% increase in the
solar constant. Their implied climate sensitivity is 0.3 to 0.9 °K per W/m^2. Note
that these sensitivities are in units of an "effective solar constant" obtained by
dividing the solar constant by 4 to account for the difference between Earth's
absorbing and radiating areas. The Global Climate Models (GCMs) use about
1 °K per W/m^2 for the climate sensitivity to solar input variations, requiring
0.4–1.5% solar variability, which is outside the range of solar variations measured during the past decade. From an energy standpoint, a factor of at least 10
is missing in known solar variations that would produce the observed climate
fluctuations. Many meteorologists and climatologists remain skeptical about
whether solar activity variations are significant at all. Regardless of the past
influence of solar activity, its present significance, relative to anthropogenic
effects, would likely differ from the historical record since man-made influences were nonexistent before modern civilization.

At least four responses to this situation are possible: (1) agreeing with the
skeptics and denying that solar activity does not significantly affect climate; (2)
considering the possibility that long-term solar-constant variations are significantly larger than the variations seen during the last decade; (3) believing that
the Earth's atmosphere is more sensitive to solar influences than our understanding of climate behavior suggests; and (4) evaluating the most important
distribution of Earth's incoming solar radiation.

Although it is not clear which view will be adopted, we shall now examine the fourth approach, choosing solar ultraviolet (UV) variations as a source for natural solar climate influences and showing that the Earth's climate may be highly sensitive to this energy input. We also discuss how the choice of boundary conditions may affect the solution.

Solar Ultraviolet Variations as a Source of Climate Change

In the past, solar ultraviolet (UV) variations have not been considered a major source of climate change. In 1991 Crowley and North believed the impact of UV fluctuations to be "problematical" compared with the total irradiance variations. Heath and Thekaekara in 1977 thought increases in the solar UV were generally compensated for by decreases at other wavelengths, which today we know is not true. In 1992 Lacis and Carlson stated: "An increase in solar ultraviolet radiation, which is absorbed primarily in the stratosphere, would cause local stratospheric heating with the excess thermal energy being radiated directly to space. This produces no net forcing at the tropopause, hence no increase in surface temperature." In 1980 Borucki and his colleagues considered the influence of a varying UV on various time scales. Although their correlation was interesting, they found it would "need to be reversed" to explain known climate changes. One major and well-known effect of the UV radiation is on the upper atmosphere ozone. UV variations are an excellent candidate for solar variability influences on climate not only because solar spectral irradiance fluctuations are proportionally larger at short wavelengths but also because they carry a significant fraction of the total solar energy variability (about 20% below 300 nm according to Lean in 1991). Ramanathan and Coakley also reviewed the field of radiation influences in climate modeling in 1978.

Consider the following problem concerning UV influences on climate: if the UV increases by a certain amount while the solar constant remains fixed, what will be the impact on Earth's climate (that is, what will the atmospheric temperature structure be)? The solar constant in this scenario is chosen not to increase so that one can examine the spectral influence variations separately from solar-constant variation effects. Additionally, when the sun's activity increases, both sunspots and faculae increase. Faculae represent bright photospheric regions, and sunspots represent dark regions. Sunspots and faculae are two phenomena with opposing and roughly equal effects on the solar constant, making the solar-constant changes smaller than the spectral irradiance changes. There is a reason the activity-related solar-constant effects are small. Solar activity influences can be thought of as a local magnetic inhibition of energy transport below sunspots, although flows play a role in this inhibition process, so that the convection-zone transport of energy does not supply the sunspots with energy. The global "river" of energy flowing out of the sun, however, cannot easily be dammed within the thin layers below the photosphere, and thus the energy must continue to flow outward despite the presence of the sunspot inhibitors. Faculae are the sites that allow this river of energy to continue to flow outward. Faculae resemble sunspots as magnetic wells; however, faculae are small rather than large. This shape change is significant because it allows their thin hot walls to radiate excessively at oblique heliographic angles. This effect allows the river of solar energy flux to emanate from the photosphere. Thus, solar activity only marginally affects the solar constant. In addition, the extra "churning" associated with the resultant flows induced by the active regions appears to stir-up the hotter subphotospheric layers and actually allows the sun to radiate somewhat excessively when solar activity increases.

Active regions produce a larger effect on the sun's spectral output than they do on the solar-constant effects. Faculae shine primarily in the blue and UV zones, while sunspots provide a general decrease in solar output at all spectral wavelengths as a portion of the sun's photosphere darkens. When both these photospheric manifestations of solar activity occur, the blue and UV wavelengths increase and the redder wavelengths decrease.

To create a relatively simple model one can remove the solar-constant changes by allowing the UV radiation energy to increase and the visible light energy to decrease by an equal amount. Given these perturbations to the visible and UV fluxes, how will Earth's atmosphere respond? Manabe and Strickler (M&S) (1964) have obtained a solution to this problem, shown in Figure 4.8. This is basically a one-dimensional model of the climate which has advantages over the three-dimensional general circulation models, in that a steady state

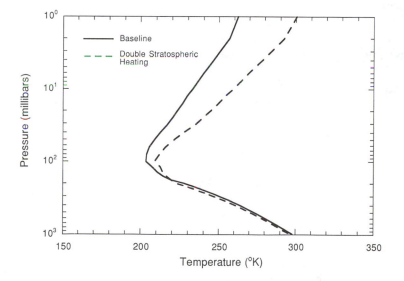

FIGURE 4.8 Two atmospheric vertical temperature profiles, after Manabe and Strickler (1964). The curve labeled Baseline shows the temperature profile of the Earth's atmosphere normally, and the dashed curve shows the modified profile, based upon a doubling of stratospheric heating (solar UV energy). The doubled energy is more than 10 times too large to mimic the solar changes, but shows the effect more clearly.

solution is more easily obtained in this model. Climate modelers have traditionally relied upon both general circulation models (GCMs) and one-dimensional climate models (M&S model). To handle weak forcings such as solar influences, the one-dimensional models are better since it is too difficult to reach equilibrium with the modified boundary conditions using the GCMs.

Because solar energy seems insufficient to create significant climate warming, many advocates of solar influences on weather and climate have searched for a "trigger mechanism" capable of releasing significant quantities of energy quickly, amplifying a small solar influence so that it overrides the meteorological influences. This type of climate change mechanism has not been sufficiently considered.

Empirical and Theoretical Approaches to Understanding Climate Change

A true understanding of climate change requires field experiments, empirical studies, and theoretical models. A purely empirical approach can lead potentially to the discovery of relationships between different parameters, but will not explain these connections. Without some theoretical framework, the empirical results can easily be dismissed as coincidences. If, on the other hand, a connection is found empirically between two phenomena that cannot be explained theoretically, the theory may need to be improved.

Theoretical models provide a guide for empirical studies. The models may predict previously unsuspected relationships that experiments and empirical studies can either confirm or disprove. Advancement requires the interaction between theory, experiment, and empirical models.

The next few chapters will consider some empirical studies of climate change within the context of the solar and climate theories and observations. Many scientists would dismiss these studies, preferring to approach the problem from a purely theoretical viewpoint. Theoretical modeling imposes solar-irradiance variations upon a general circulation model. A final and definitive conclusion about the importance or unimportance of considering solar variations is taken from the model output. Although this approach is probably deemed appropriate, danger lurks in the inadequacies of the general circulation models with 100,000 or more lines of code that may not include all physical processes or be error-free. Many shortcuts, primarily empirical approximations that may not always be adequate, increase computational speed. These complex models contain so many factors, that achieving "correct" results means they must be tuned. Considerable caution should be taken in interpreting their answers. A few of the many potential problems with these models are listed here:

- Instead of calculating the vertical transport of energy, the models assume it is efficient enough to establish an adiabatic lapse rate, as discussed in the last section.
- Rather than predicting cloud amounts based on physical laws, the models instead use empirical parameterization that may be applica-

ble to short-range weather predictions but are not adequate for studies of climate change.

- The models do not achieve energy balance throughout the atmosphere, but use ad hoc corrections to make the models function and
provide a solution.
- The models cannot calculate Earth's mean surface temperature to
better than 0.5 °C (and until recently to within 5 °C), yet these same
data are used to predict changes in the means of a fraction of degree.
- The models omit physical processes such as a change in the ocean-
surface albedo as a function of wind speed, no doubt because including such processes would be computationally intensive.
- The indirect influences of solar activity upon upper atmospheric
opacities, through the concomitant changes in thermodynamic state
variables, need to be considered.

The above criticisms could be extended over many pages. These criticisms
merely suggest that even after more than 20 years of work, the GCMs are still
in an early stage of development. Modelers are aware of many shortcomings
and are continually striving for improvement. Meanwhile, climatologists, policy
makers, and others should remain cautious and skeptical about these models.
Not only are more improvements required, but also the results from the models
need to be balanced with measurements and empirical studies.

Searching for Empirical Solar Influences

If we postulate a solar, influence on climate, such as an 11 year variation in
solar irradiance (see Figure 3.12), where can we expect to see a solar-induced
climate signal? In general, we expect to find a weak signal in a noisy time
series. The following are some useful guidelines:

- Accurately and self-consistently measure the variables used. Data inhomogeneities only obscure any possible signal.
- To improve the signal-to-noise ratios, use low-spatial-variability climate variables. Temperature, for instance, is more promising than
precipitation.
- Since long time spans greatly aid in detecting a signal, concentrate
on large-scale spatial variations. Use data binning in monthly or
yearly blocks when searching for weak signals.
- Some regions may have stronger signals than others. Surface temperatures over continents will probably vary more than over oceans.
Because of the spectral variations in the solar output, changes high
in the atmosphere will be larger than changes lower down.
- Some changes in the means are so small it is worthwhile seeking
signals elsewhere. Changes in the frequency of extreme weather
events are particularly amplified for small changes in the mean, making these changes useful tools in searching for solar forcing signals.

Summary

Even for a measure like temperature, reconstructing past climate records is not straightforward. Many possible errors lead to a false picture of climatic change. Unfortunately, long-term trends of primary interest are the most likely source of errors. Mistakes occur because much of the Earth is not measured, temperature-measuring techniques change over time, shelter types vary by place and time, local urban heat islands develop, combining measurements from an unequally spaced grid of stations is complicated, and so forth. Despite these problems, we probably have a reasonable reconstruction of Northern Hemisphere temperature trends since 1880. Nonetheless, we should be skeptical about the upward trend in temperature, equal to about a 0.7 °K warming. Perhaps 0.1 °K is due to spatial averaging problems, and another 0.1 to 0.2 °K could arise from urban heat island effects, so only 0.4 °K may be real. Chapter 10 covers the amount of warming caused by the sun, while chapter 11 explores the degree of warming caused by mankind's activities, known as the carbon dioxide greenhouse effect. Even when reliable climate records are available, improper use of statistics can lead to incorrect conclusions. Some of the statistical problems are mentioned in the text and in Appendix 3.

The physical basis for solar influences leading to climatic change comes from satellite measurements which reveal that the sun has at least an 11-year solar-irradiance cycle with a value of about 0.15%. If the Earth behaves as a blackbody, this could lead to a 0.11 °K cycle in the Earth's temperature, but a larger figure is possible because of such enhancing effects as the UV indirectly affecting atmospheric infrared opacities. The climate at any one location or region is very noisy, as weather variations predominate. Single station daily records are the worst place to look for sun/climate relationships. The best chance of observing a solar effect is using weather conditions averaged over a year or longer on a regional, hemispheric, and global basis. The next four chapters cover some proposed sun/climate relationships. Temperature, rainfall, storms, and the biota will be examined to see what evidence they provide in support of 11-year signals that could be attributed to changes in the sun's brightness.

5. *Temperature*

Theoretical Expectations

How the bulk of the sun's energy variations, which arise from the solar-irradiance changes, contribute to terrestrial changes will be the subject of this and the next few chapters. Although it's a simplification, the hypothesis that only the largest solar activity–related energy contributors to the Earth's atmosphere need to be considered allows us to ignore such far-flung ideas as the influence of sector boundary crossings, cosmic rays, and other less energetic phenomena. If the Foukal and Lean model of solar irradiance (see Figure 3.14) is a close approximation of known solar behavior on the 11-year time scale, what are the climatic consequences of these variations? We can ignore all the other proposed solar cycles ranging from 6 to 7 days to hundreds or thousands of years. Although some of these other proposed solar cycles may be real, we will not indulge in cyclomania here.

North and his colleagues in 1983 developed a theoretical energy-balance model with a geographical distribution of land and ocean. Testing the model to see how well it reproduced the observed annual cycle of temperatures revealed satisfactory agreement, so North et al. subjected their model to other cyclic solar forcings. Figure 5.1 shows a 10-year cycle imposed on the Earth. Climatologists were not too surprised by the conclusions, as the solar-irradiance and temperature changes are nearly in phase because Earth's time constant (about 3–5 years) is less than the imposed cycle time. The response over the land is greater than that over the oceans because the land holds less heat than water does and responds more strongly. The amplitude of the temperature variations is very small, no larger than about 0.11 °C. The maximum response is also centered near Arabia.

Not everyone agrees that Earth's temperature response to solar cycle changes will follow the scenario shown in Figure 5.1. In the last chapter, we

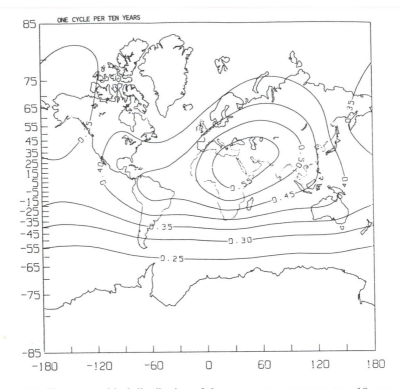

FIGURE 5.1 The geographical distribution of the temperature response to a 10-year solar forcing. To convert the numbers to degrees corresponding to the Nimbus-7 observations, the contour numbers should be multiplied by about 0.2. Thus, the 0.55 °C, contour should be 0.11 °C, and so forth. (From North et al., 1983, with permission.)

described a novel approach suggesting the possibility of a larger response to solar activity than North's and similar models provide. Kim in 1994 argues that tropical ocean waters, about 30° north or south of the equator, are the center of response to solar variations, (personal communication). At these same latitudes over land, the desert regions might have a maximum response. Kim suggests that since both ocean and desert regions remain mostly free from clouds, they will respond directly to solar variations. Despite these arguments, when examining empirical studies in this chapter we will use North's theoretical response.

Early Hints of a Solar-Variability, Earth-Temperature Connection: 1610 to 1859

Before examining the more modern studies of Earth's temperature and solar variability, it is worthwhile considering some earlier ideas. In the 1600s, there were several speculations about sunspots and climate. The first modern writing might be by Antonii Mariae Schyrlei de Rheita of Antwerp, who in 1645 suggested that an increase in sunspots as associated with stormier and colder weather.

In *Almagestrum Novum* published in 1651, Riccioli states that colder temperatures are associated with more sunspots, basing his comments on observations. In *Iter Exstaticum Coeleste* in 1671, Kircher may be repeating Riccioli's conclusions, but he too states that more sunspots mean colder and rainier conditions. On the other hand, in the book *Astrometeorologica or Aphorisms and Discourses of the Bodies Celestial, Their Natures and Influences* published in 1686 skeptic John Goad says: "the Sun labors and is disturbed at such times, as the learned writers of the macular observations conclude, Scheiner and Hevelius. All that I have to say is, this inquietation comes from the heavens. In the body of the celestial sphere, one part affects another." At first glance, this quotation appears to support the idea that sunspots influence weather, but Goad later says that during the existence of any one sunspot, many types of weather occur, so sunspots cannot be influencing weather.

In 1676 Robert Hooke arrived at the opposite conclusion, closer to the modern understanding of the sun. In a two-paragraph discussion of the influence of sunspots on weather, Hooke concludes, "it seems worthy remarking that the greatest heat that hath been in the Air this year was on the day of June when the first spot was near the middle of the Sun." He does find contradictory evidence and cautiously states, "at least this Hint may deserve some further Inquiry. . . . Possibly the appearances of the Spots may serve to predict the future constitution of the weather."

In 1729 J. B. Wiedenburg of Helmstadt associated cold and stormy weather with more sunspots and aurorae. His ideas were widely reviewed in the contemporary scientific literature, but this is the last discussion of the issue we can find until Sir William Herschel again raised the topic in 1801. In his observations in the *Philosophical Transactions* of 1801, Herschel notes that high wheat prices, indicating a scarcity of wheat, occur when sunspots are few. Because the amount of wheat depends in some manner on the amount of sunlight, he concludes that lower solar activity means less light and therefore less wheat. In other words, higher solar activity and Earth temperatures go together. This conclusion is exactly the opposite of all his predecessors. After his 1801 publication, Herschel kept observing both the sun and weather and recorded some of his thoughts on this subject in his notebooks. He refers to sunspots as "openings." On December 13, 1801, in the Additional Observations, he wrote: "It seems from the present state of the Sun that the influence of the openings, supposing them to be symptoms of a copious emission of solar rays is not immediate. There having been strong frosts." Here, Herschel seems to admit that his observations and theory do not match, and he believes sunspots are warmer than the rest of the sun, which we now know is false. On January 23, 1809, Herschel expressed further doubts, writing: "If the openings on the solar disk have any effect on the quantity of heat emitted, it should not remain many days so cold. It is however, uncertain how long should be allowed for the effect becoming sensible." Such anecdotal evidence suggests that Herschel associated colder weather with more sunspots.

In 1823 Flaugergues said that the weather was warm when no sunspots were present and cool when there were many spots. At the time, Flaugergues had been observing sunspots for 35 years. In 1826 Gruithuisen concluded: "Settled fine weather occurs on the earth, when on the Sun the variable weather (that is, sunspot formation) ceases. Great spots call forth on the earth variable weather differing greatly in different localities. The more scattered the spots occur, the less does the temperature of the earth's atmosphere rise since only spot groups or great spots send forth more heat." Gruithuisen implies that higher temperatures occur with fewer sunspots, agreeing with all earlier observers except Herschel.

In 1860 Robert Greg tried to interest more scientists in sun/weather relationships. At the conclusion of a letter he says: "The opinions of philosophers differ respecting the influence of a paucity or an abundance of solar spots upon the temperature and seasons of the earth; the probability is, there is simply a general disturbance, arising from an increase of (solar) magnetic influence, which may produce greater heat and dryness in one part of the globe, and more cold and rain in other parts. I have merely hazarded these few remarks in the hope of drawing further attention to such interesting and important topics." Despite this plea for more research, the next 10 years saw little discussion of the topic. When C. Piazzi Smyth raised the issue again in 1870, the furor and controversy resumed in earnest.

C. Piazzi Smyth and the Start of a Controversy

C. Piazzi Smyth's solar radiation measurements used the Edinburgh boreholes. Smyth continued analyzing these borehole measurements, and by 1870 he had prepared another paper on the subject communicated to the Royal Society called "On supra-annual cycles of temperatures in the Earth's surface-crust." Using inclusive data from 1837 to 1869, Smyth says there are three cycles longer than 1 year and that the strongest of these was 11.1 years. He concluded that the cycles in rock temperatures were probably solar induced, but indirectly through a weather influence.

Soon after Smyth, E. J. Stone, the recently appointed Astronomer Royal of the Cape of Good Hope, submitted another paper in 1871 on this same subject entitled "On an approximately decennial variation of the temperature at the Observatory at the Cape of Good Hope between the years 1841 and 1870, viewed in connexion with the variation of the solar spots." The controversy between Smyth and Stone is mainly one of priority. Who made the discovery? Both authors actually found that more sunspots mean lower temperatures, a finding which had already appeared in the German and French scientific literature. Neither Smyth nor Stone seems aware of these earlier works. In his paper, Stone says: "I may mention that I had not the slightest expectation, on first laying down the curves, of any sensible agreement resulting, but that I now consider the agreement too close to be a matter of chance." Smyth questions whether this was actually the case because his own paper, submitted to a secret committee of the Royal Society, was held up for 9 months. At this time, Stone was in or near London and he could have heard about or had some knowledge of Smyth's results. To make a judgment after the passage of 125 years is impossible, but the controversy often attracts attention to a topic that might otherwise be forgotten. The question of priority was not the only controversy, as the reviewers of Smyth's work said he "was not to be allowed to compare the Edinburgh mean annual temperatures with sun-spot observations." This reviewer's sentiment would be repeated many times during the years to come.

The actual findings of both Smyth and Stone are less important than the effects their papers had on others. For the first time, the English-speaking scientific community was becoming aware of the possibility of sun/climate connections. In the 10 years before the works by Smyth and Stone, only about three papers even commented on the topic. In the following decade, about 100 papers appeared. By bringing the topic into common view, Smyth and Stone provided serious consideration of sun/climate relationships.

The Work of Wladmir Koppen

Of the 100 or so papers that appeared during the 1870s which discussed sun and climate, Wladmir Koppen's paper stands out. The article was written in German and appeared in *Zeitschrift der Osterreichischen Gesellschraft für Meteorologie* in 1873: an English translation of the paper's title is "On the many year periods of weather, especially about the eleven-year period of temperature." This classic paper is the most comprehensive study of the topic up to that time. Koppen concluded that the sun does indeed cause a periodicity in the Earth's surface, not only in many locations but also on a hemispheric and global scale. His four major conclusions were:

- In the tropics, the maximum temperature occurs about a year before the sunspot minimum.
- In the mid-latitudes, the maximum temperature occurs from zero to 3 years after the sunspot minimum.
- The shape of the temperature cycles is the same as the shape of the sunspot cycles—that is, a rapid decrease followed by a slower rise.
- The clearest signal is a negative correlation between temperature and sunspot number from 1815 to 1854. From 1777 to 1790 a positive correlation exists.

In 1914 Koppen repeated and updated his earlier study. This second article examines a full century of data and again concludes that an 11 year solar cycle exists in Earth's surface temperatures. Examining Koppen's conclusions, the first question one might pose is whether he could calculate temperature variations on a hemispheric scale. The answer is yes. Figure 5.2 shows five different temperature reconstructions for the Northern Hemisphere. Koppen's values end in 1910, but where they overlap modern reconstructions, they essentially agree.

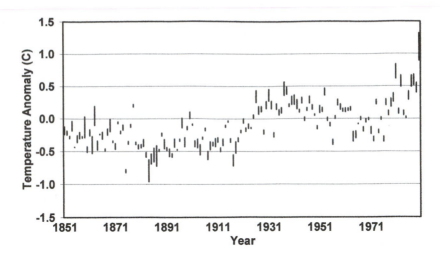

FIGURE 5.2 Variations of the Northern Hemisphere surface-land temperatures during the last century. Each year is represented by a bar showing the 1 SD uncertainty in the annual mean temperature for that year. The three temperature reconstructions of Viinikov, the Goddard Institute for Space Studies (GISS), and the Climate Research Unit (CRU) are used. Although they all use basically the same raw data, different analysis techniques give different final answers, which give rise to uncertainties in the final temperature reconstructions. All the reconstructions have the same general trends with warming to about 1940, followed by 20–25 years of cooling, with a resumed warming to the present.

On this basis, we should give some credence to Koppen's reconstruction for the years before 1850.

Now let us focus on Koppen's temperature reconstruction. Figures 5.3 and 5.4 suggest that a solar signal is evident in the temperatures. Figure 5.3 shows the Koppen temperature anomalies and the inverted Wolf Sunspot Numbers from 1811 to 1910. Both curves parallel each other quite well, suggesting a cause-and-effect relationships, which appears stronger before 1870 than after 1870. Just before 1890, the relationship almost seems to disappear. Another way to examine this data set is to use a scatter plot, with temperature anomalies plotted as a function of the Wolf Sunspot Number (Figure 5.4). This clearly shows that as the Wolf Sunspot Number increases, the temperatures decrease. Nearly all the investigators before 1910 got essentially the same results. A linear regression fit (Eq. 1) through the data in Figure 5.4 gives

$$T = (-0.21 \pm 0.23) - (0.0035 \pm 0.0007)R_z \qquad (6)$$

in which T is the temperature anomaly and R_z is the Wolf Sunspot Number. This result implies that about 22% of the variance is attributable to the sun.

At this point, we need to consider why these empirical results are so differ-

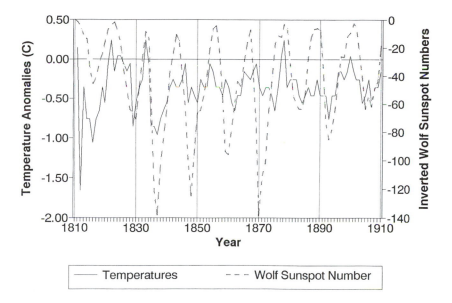

FIGURE 5.3 The Koppen temperature anomalies and the Wolf Sunspot Numbers plotted simultaneously. The Wolf Sunspot Numbers are inverted to show how well the two curves track. Lower solar activity is associated with higher temperatures, and higher solar activity is associated with lower temperatures. Many authors confirmed these conclusions.

ent from the expected theoretical results, given the known variations in solar irradiance for 1978–1993 and North's climate model results.

A Paradox of Sun/Climate Relationships: Why Are the Empirical and Theoretical Results Contradictory?

Over the last solar cycle, the irradiance increased by about 0.15% from solar minimum to maximum. From North's model, we would expect about a 0.05 °C change in global mean temperature. The Wolf Sunspot Numbers changed by about 150, so we could approximate the temperature variations (T) as equal to about $0.0003 \, R_z$. In contrast to equation 6 in this chapter, this number has the opposite sign and a magnitude only about one-tenth the empirical result. This paradoxical outcome suggests several possibilities, among others:

1. Koppen's empirical correlation arose by chance. The temperature variations are random or controlled by some other process, such as volcanic eruptions. Or that there are common periodicities or persistencies for temporal variations on the Sun and Earth's atmosphere, making chance correlations easier to happen.

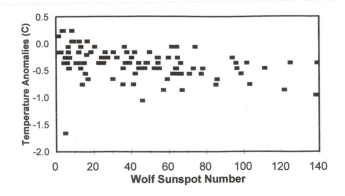

FIGURE 5.4 A scatter diagram plotting the Koppen temperature anomalies versus the Wolf Sunspot Numbers. A definite linear relationship seems apparent.

2. Before about 1920, the solar irradiance decreased rather than increased with solar activity. Perhaps, as Lockwood's diagram (see figure 3.21) suggests, the sun is an unusual star, sometimes having irradiances positively correlated with solar activity and sometimes negatively.

3. The Earth's climate is more sensitive to solar variations when it is cooler than when it is warmer, as it is today. Most climatologists would probably reject this idea and say the apparent large sun/climate sensitivity implied by Koppen and others is further evidence that the findings are coincidental.

To support the idea that we are dealing with a coincidence, Figure 5.5 repeats Figure 5.4 using temperature anomalies after 1910. This scatter plot shows that any correlation existing before 1910 has disappeared in the modern data. Even a sign reversal is indicated in the relationship. This sign reversal is most evident using data from the last few solar cycles. After examining Figure 5.5, modern scientists would conclude there is no sun/climate relationship. But after looking at Figure 5.4, the scientists of the previous century would conclude there is a definite relationship. Arthur Schuster very definitely stated this older conclusion in 1884 before the British Association when he said: "There can be no longer any doubt that during about four sun-spot periods (1810 to 1860) a most remarkable similarity existed between the curves of nearly every meteorological element which is related to temperature. This is not, in my opinion, a matter open to discussion: it is a fact."

Another Controversy or a Simple Misunderstanding?

Koppen was not the only person to find a negative correlation between Earth surface temperatures and solar activity. In the following years, many studies confirmed these results at varied locations and regions. Hahn, Fritz, Hill, Bruck-

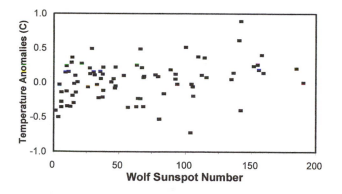

FIGURE 5.5 A scatter plot showing the post-1910 Northern Hemisphere temperature anomalies as a function of the Wolf Sunspot Numbers. No relationship is apparent, and this plot is quite different from Figure 5.4. Here an increase in sunspot numbers of 100 corresponds to a temperature increase of 0.1°C. By contrast, in Figure 5.4 the same sunspot number increase corresponds to a −0.35°C temperature drop.

ner, Blanford, Gunther, Arrhenius, Angot, Easton, Hann, Newcomb, Arctowski, Humphreys, Mielke, and Helland-Hansen and Nansen all made temperature studies between 1877 and 1920. To illustrate the consistency of the belief that the absence of sunspots caused the Earth to warm, consider the case of C. J. B. Williams. In 1883 Williams published an article in *Nature* called "On the cold in March, and absence of sunspots." This article brought an immediate rebuke claiming that Williams's sign was wrong and that a month of research is too short a time to conclude anything. Nordmann (1903) provides another example of the consistency between the different studies when he states: "The mean terrestrial temperature exhibits a period sensibly equal to that of the solar spots; the effect of spots is to diminish the mean terrestrial temperature, that is to say, the curve which represents the variations of this is parallel to the inverse curve of the frequency of solar spots." In 1928 Shaw provided a table correlating solar activity and temperature using earlier studies. We repeat Shaw's table here, listing stations from north to south (Table 5.1).

By about 1920, something strange began to happen to the correlations and 11-year cyclic temperature variations. D. Brunt (1925) looked at temperatures in Edinburgh, Stockholm, London, Berlin, Paris, and Vienna. At these six locations, which had tabulated annual means that start between 1756 and 1775 and end between 1863 and 1918, Brunt found an 11-year cycle only at Edinburgh, which had the shortest record, spanning 1764 to 1863. Others had already confirmed an 11-year cycle for this time span. Using long time series, Brunt concluded that a solar cycle signal was not evident in the other five cities.

During the next 30 years or so, the entire field of sun/climate relationships becomes rather confusing. Some authors report positive correlations, others negative, and still others no correlation at all. The results seem dependent upon the chosen geographical region, the time interval, and the analysis technique.

TABLE 5.1 Correlation between annual temperatures and sunspot numbers (adapted from Shaw, 1928; values of −.35 are approximate).

Station	Latitude	Correlation Coefficient
Yakutsk, Russia	62.1 N	−.35
Winnipeg, Canada	49.6 N	−.55
Victoria, Canada	48.3 N	−.35
Sydney, Canada	46.1 N	−.35
Batavia, Illinois	41.5 N	−.35
Tashkent, Russia	41.2 N	−.35
Baghdad, Iraq	33.1 N	−.35
Brisbane, Australia	27.3 S	−.42
Arga, India	27.2 N	−.43
Calcutta, India	22.3 N	−.44
Hong Kong	22.2 N	−.45
Bombay, India	18.6 N	−.35
Newcastle, Jamaica	18.0 N	−.35
Recife, Brazil	8.1 S	−.45
Cordoba, Argentina	30.2 S	−.35
Pelotas, Brazil	31.5 S	−.58
Santiago, Chile	33.3 S	−.35
Sydney, Australia	33.6 S	−.49
Auckland, New Zealand	36.5 S	+.27

From Shaw 1928.

How difficulty it was to understand what was going on is demonstrated by the introduction of more elaborate hypotheses, often an indication that a paradigm is in trouble. For example, Clayton in 1943 stated that "the increase of temperature at the Earth's surface with increased sunspots takes place chiefly over continental masses in low latitudes and especially desert regions." This conclusion is exactly the opposite of Koppen's and those of other earlier authors. Were these earlier analyses wrong? After Brunt's classic study in 1925, there are fewer solar-activity/Earth-temperature studies, particularly in English. Most subsequent studies appeared in a number of European languages and often reached different conclusions.

For most scientists, all this confusion caused the sun/climate field to fall into disrepute. In 1965 R.A. Craig wrote: "Fifteen years ago [i.e., 1950], the study of Sun-weather relationships was considered by many to be an undignified pursuit for a meteorologist." In June and July 1956, a conference on sun/climate relationships held in Boulder, Colorado, recommended conducting more studies and emphasizing physical mechanisms, despite disappointing findings up to that time. In contrast, the American Meteorological Society (AMS) stated in 1957 that "Some few results [in Sun/weather relationships] have been obtained that are highly suggestive. But none of these studies have produced any conclusive evidence that relationships do indeed exist." AMS researchers seem to imply that further studies might prove fruitless unless the physical basis for the results were better understood.

A few of this era's scientists recognized that one problem in the sun/ weather field was the reversal of correlations. For example, Clayton in 1940 recognized a sign reversal in the apparent dependence of water levels in Lake Victoria after 1920. In England, E. N. Lawrence recognized a similar sign reversal for English temperatures around 1930, shown in Figure 5.6.

Similar sign reversals were occurring in many locations and to all kinds of meteorological variables including rainfall, lake levels, Greek etesian winds, locations of Icelandic lows, and the number of Indian monsoons. Figure 5.7 reproduces the Herman and Goldberg (1978) summary of some of these correlation reversals and failures.

These sign reversals have several significant implications. First, without regard to this effect, a simple power spectrum analysis will be very sensitive to the interval selected. This could explain why some investigators find marked power at 11 years in their spectra while other investigators using the same variable and location fail to find a signal. Different time intervals yield different

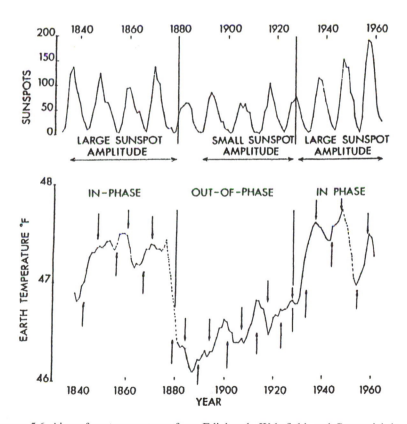

FIGURE 5.6 Air-surface temperatures from Edinburgh, Wakefield, and Greenwich in Great Britain shown with Wolf Sunspot Numbers. Temperatures appear to be out of phase with solar activity from 1880 to 1930, but in phase for other years. (Adapted from Lawrence, 1965.)

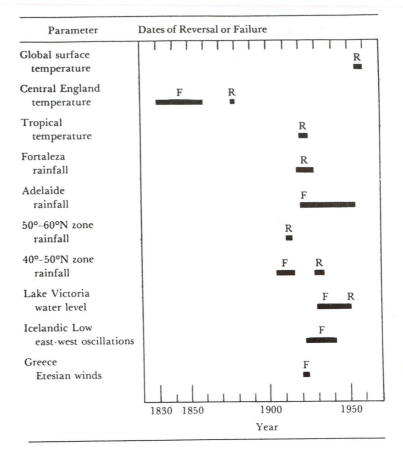

Parameter	Dates of Reversal or Failure
Global surface temperature	R
Central England temperature	F R
Tropical temperature	R
Fortaleza rainfall	R
Adelaide rainfall	F
50°-60°N zone rainfall	R
40°-50°N zone rainfall	F R
Lake Victoria water level	F R
Icelandic Low east-west oscillations	F
Greece Etesian winds	F

1830 1850   1900   1950

Year

FIGURE 5.7 Some correlation reversals (R) and failures (F) for selected meteorological variables. Additional correlation reversals exist which also cluster around 1920. (From Goldberg, 1978, with permission.)

conclusions. Modern spectral analysis techniques such as wavelet analysis giving spectra about a specified window (i.e., interval) would permit tracking the appearance and disappearance of solar cycle signals. While this approach has not yet been adopted in the field of solar cycle signal analysis, doing so could clarify much of the confusion still existing today. Contradictory conclusions using values of correlations can perhaps also be explained by these phase-reversal effects.

Many correlation reversals occurred from around 1920 to 1930. Is there a physical explanation for this effect? Lawrence in 1965 suggested that cycles with low levels of activity tend to have negative correlations with sunspot numbers, while stronger cycles have positive correlations. Could weak solar-cycle sunspot blocking dominate over facular emission? Could facular temperatures be lower during these weak cycles, allowing sunspot blocking to predominate? An argument somewhat against a physical mechanism states that not all the phase reversals and failures are simultaneous. Figure 5.7 reveals that eight of 10 reversals and failures occurred around 1920. According to Lawrence (Figure 5.6), the same effect is evident in England, despite its failure to show up in Manley's temperature reconstruction. Around 1920 these and other variables all change sign, implying a real effect rather than a statistical artifact.

The temperature sign reversal was so strong that by the time J. W. King published an article on sun/climate relationships in 1974, minor smoothing showed that the positive London temperature/solar activity correlation could be seen in the data (see Figure 5.8). Many other locations worldwide showed similar, apparent temperature dependencies on solar activity. Studies of these phenomena by Robert Currie and by Robert Lee occupy the next two subsections.

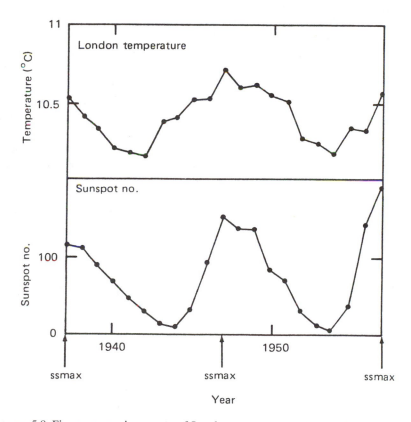

FIGURE 5.8 Five-year running means of London temperatures plotted with the sunspot numbers in the lower panel (adapted from King et al., 1974). Note the strong positive correlation here that is in agreement with the results shown in Figure 5.6, but is quite different from Koppen's conclusion a century earlier.

Robert Currie Reexamines the Evidence

Before 1974 several attempts were made to look for an 11-year temperature data signal using the Fourier analysis technique. It is difficult to use Fourier analysis to detect weak signals. In 1959 H. E. Landsberg and colleagues found an 11-year signal in the temperature records at Woodstock, Maryland. During 1970 Dehsara and Cehak looked at spectra for 92 locations worldwide and found 14 peaks from 9 to 11 years. Their results produced no reasonable geographical pattern and displayed numerous other peaks as well.

In the late 1960s and early 1970s, J. P. Burg developed a new method of calculating spectra known as the maximum entropy method (MEM) or maximum entropy spectral analysis (MESA), a very powerful technique for detecting weak signals in noisy data. This method is ideal for seeking an 11-year cycle in noisy temperature records. Robert G. Currie of South Africa recognized this and began a systematic investigation of temperature using MESA. In 1974 he published an analysis of 226 stations worldwide. Currie found a 10.6-year signal cycle at many locations. For 78 North American stations, he found a 10.5-year cycle with an amplitude of 0.27 °C. For this interval that Currie selected, the solar cycle length was 10.7 years, so his results appear to show that the sun is influencing climate. He achieved these results-even though no special effort was made to remove inhomogeneities in the temperature records caused by the various effects shown in Figure 4.1. The value of 0.27 °C is larger than the 0.08 to 0.11 °C values predicted by North's model (Figure 5.1) for a 0.15% solar-irradiance amplitude. It is possible that the solar irradiance's 11-year variation averaged about 0.43%, but given Foukal and Lean's model (see Figure 3.14), such a possibility seems unlikely. A second possibility is that climate is about three times more sensitive to a solar forcing than North's climate model indicates.

In 1979 Currie did a more detailed study of North American stations, whose records are often longer and more homogeneous than those in many other locations. Currie found a solar cycle in the northeast quadrant of North America with an amplitude of 0.29 °C and a 10.7-year period. He also found a signal in one-third of 44 European stations, but the distribution made no geographical sense. This confirms an earlier conclusion by Dehsara and Cehak, who also found that nearby stations often give different results. These studies seem to suggest that many locations had unrecognized inhomogeneities or that local geographical effects are masking any regional effects.

Currie reexamined the U.S. temperature records again in 1993. These new records were improved versions of the older records with more inhomogeneities removed. Looking at 1,197 locations, Currie reached two important conclusions. First, he found cyclic variations in temperature at nearly every location, instead of just the northeast. Power spectra showed peaks at 10.4 and 18.8 years (Figure 5.9). The first peak arose from solar-irradiance variations and the second possibly arose from lunar tidal effects. The second major finding was that east of the Rocky Mountains these cyclic variations are in phase with the solar-irradiance variations, while west of the Rocky Mountains they were 180°

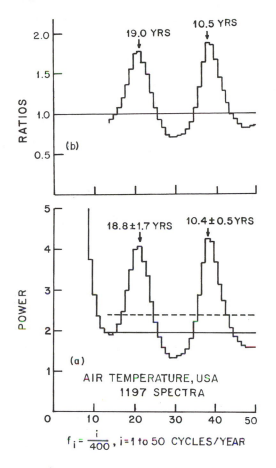

FIGURE 5.9 The composite power spectra for 1,197 locations in the contiguous United States since 1900. The predominant 10.4-year peak can be identified with solar forcing. The 18.6-year peak is caused by the Saros lunar cycle that modulates tidal forcing. (From Currie, 1993, with permission.)

out of phase (Figure 5.10). Currie argues that large atmospheric standing waves keep some regions in phase and some regions 180° out of phase with solar variations for many decades. Currie believes there may be three global dividing lines like the one seen in Figure 5.10 along the Rocky Mountains. The dividing lines have very sharp widths of only about 20 miles. These standing wave like patterns maintain themselves for many decades before breaking down and rearranging themselves into a new pattern.

Many studies show a measurable solar influence on U.S. temperatures. Although weak and difficult to detect, the signal still appears stronger than climate models predict. Currie has found similar effects around the world, wherever reliable observations can be found. Substantiation of these conclusions has now come from satellite measurements made over the last solar cycle.

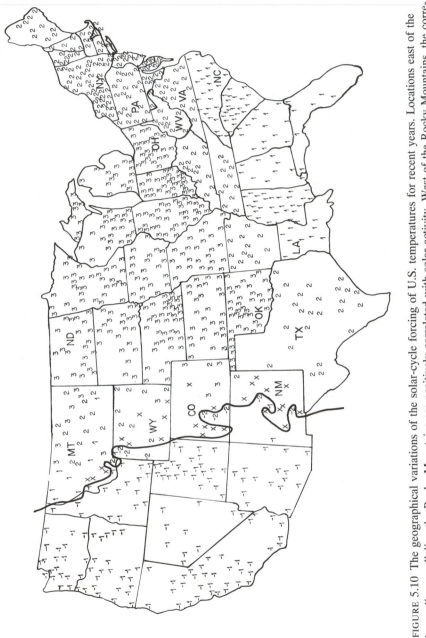

FIGURE 5.10 The geographical variations of the solar-cycle forcing of U.S. temperatures for recent years. Locations east of the heavy line paralleling the Rocky Mountains are positively correlated with solar activity. West of the Rocky Mountains, the correlation is negative. Numbers indicate relative strength of the correlations. (From Currie, 1993, with permission.)

The Last Solar Cycle

Obtaining long-term self-consistent temperature measurements is difficult. Recently, however, two new satellite techniques have been developed which allow measurements of the global temperature. One method uses microwaves and the other uses thermal radiation. Since 1978 both techniques have provided nearly the same results. While both techniques have advantages and disadvantages, a major advantage of satellite observations compared with surface-air temperature measurements is their geographical coverage. Satellites provide nearly uniform global coverage, while surface measurements tend to be concentrated over continents and in populated regions.

Robert B. Lee of NASA's Langley Research Center has been actively pursuing temperature measurements using thermal radiation and comparing them with the solar-irradiance measurements. These measurements show a 0.3 to 0.6 °C cooling of the Earth from 1979 to 1985 followed by a 0.2 to 0.3 °C warming to 1989. Part of an 11-year oscillation in global temperatures seems to have been captured in the satellite measurements. According to Lee, if solar forcing is the only cause of these variations, temperatures respond with an amplitude five times greater than climate models predict. Recall that Currie's North American measurements had an amplitude three times greater than model predictions. Lee's larger response may have occurred because the 1982 El Chichon eruption could have cooled the Earth in 1983–1984, making the amplitude of the temperature variations larger. Lee's results are summarized in Figure 5.11.

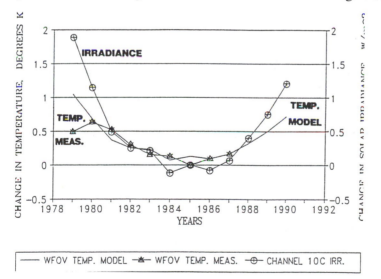

FIGURE 5.11 The yearly mean Nimbus-7 solar total irradiance values are plotted along with the temperatures deduced from thermal radiation measurements on board Nimbus-7 (the wide-field-of-view [WFOV] temperature measurements). Both curves parallel each other, suggesting a cause-and-effect relationship. (From Lee, 1992, with permission.)

Quasi-Biennial Oscillation and Polar Stratospheric Temperatures

Perhaps the most famous connection between temperature and solar activity that has withstood the test of time is one discovered by Karen Labitze and Harry van Loon. In 1988 Labitze and van Loon found that the high-altitude temperatures in the northern polar atmosphere are correlated with solar activity during the east phase of the quasi-biennial oscillation (QBO) in the tropics, a roughly 2-year period during which the tropical stratospheric winds blow eastward about half the time and westward the other half. Examining all the polar temperature data and correlating them with solar activity reveals no apparent relationship. However, dividing the data into two sets called the east phase and west phase after the wind directions in the QBO produces some unexpected results. During the east phase no correlation is evident, but during the west phase a strong correlation has been known to exist since observations began in the early 1950s. Figure 5.12 shows these effects graphically. No explanation currently exists for this relationship. Perhaps a radiative forcing mechanism is involved, or it may be something else entirely. The correlation continues to the present.

Some Final Comments

During the past 100 years, many studies have attempted to correlate temperature measurements and solar activity. The few studies mentioned here concentrate mostly on regional and global results. Individual stations are very noisy and prone to inhomogeneities in their measurements, so we have avoided reviewing the large number of site-specific temperature studies. We have also concentrated on the 11-year cyclic variation because recent satellite measurements of solar irradiance suggest the existence of an 11-year solar-irradiance cycle. Given this probable forcing in the past, one seeks the temperature response.

Loosening these guidelines somewhat introduces many contradictory and null results. Many locations demonstrate no apparent 11-year temperature cycle. For example, Shaw in 1965 found no evidence for 11-year cycles in New York City, the Netherlands, and England, and from this limited data set concluded that a sun/climate connection is unlikely. On the other hand, some authors who examined Berlin air temperatures found an 11-year cycle, while others were equally certain that no such signal exists. However, the time periods studied and the analysis techniques used differed. Perhaps part of the confusion arises from the sign changes for the temperature-solar activity relationship. Koppen and others reveal the following observed relationship between temperature and solar activity:

- For 1600–1720, a negative correlation
- For 1720–1800, a positive correlation
- For 1800–1920, a negative correlation
- For 1920 to the present, a positive correlation

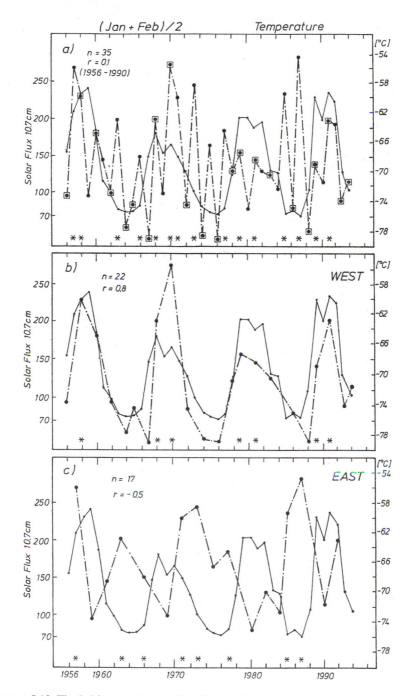

FIGURE 5.12 The Labitze-van Loon effect (from Labitze and van Loon, 1990, with permission). The 10.7-cm Radio Flux *(solid line)* closely parallels the Wolf Sunspot Numbers. The dashed lines show the polar temperature at 30 mb (about 15 miles up) for all the data, for the west phase of the quasi-biennial oscillation (QBO), and for the east phase of the QBO, respectively. During the west phase *(middle panel)*, polar stratospheric temperatures and solar activity closely parallel each other.

These sign changes are presented as global phenomena, but for the early years they rely heavily on European measurements. From recent satellite solar-irradiance measurements we expect only a positive correlation for the entire Earth. How can the negative correlations be explained? Four speculative possibilities exist.

Possibility 1: For some solar cycles, sunspot blocking dominates over facular emission, so the solar irradiance is minimized during a sunspot maximum. This causes Earth's surface temperatures to be at a minimum during the sunspot maximum, producing a negative correlation. Currently, there is no experimental justification for this possibility. Comparing the sun with solar-like stars, we find that the sun is near the borderline at which sunspot blocking dominates. Only in young stars is blocking expected to dominate, and the sun cannot be classified as young.

Possibility 2: A weak positive correlation exists at all times, but other climatic factors sometimes mask it or reverse its sign. A major explosive volcanic eruption can cause cooling for 1 to 2 years after injecting aerosols into the stratosphere. The largest volcanic eruptions occurred in 1812, 1815, 1821, 1831, 1883, 1902, and 1912, and the associated cooling would have taken place mostly near sunspot minima, leading to a positive temperature correlation with solar activity. However, the actual temperature variations were in the opposite direction, so if another climatic factor is involved in causing an apparent negative correlation of temperatures and solar activity, it is not volcanic activity but some unknown agent.

Possibility 3: Currie's view that certain Earth regions are always positively correlated with solar activity while other regions are negatively correlated is correct. Presumably, the positive regions have a greater area, so a calculated global signal will yield a positive correlation between temperature and solar activity, as shown by Lee. The earlier phase reversals reported by Koppen are caused by inadequate global sampling and mostly represent European phase changes. In short, although Koppen may have thought his global temperature representation was good, it was not.

Possibility 4: Common periodicities or persistencies exist on the Sun and Earth, enhancing chance correlations. Namely, the Earth's temperature varies naturally over a range of time periods, with some close to 11 years. This random variation is sometimes in phase with solar activity and sometimes not, leading to the correlation sign reversals.

No doubt other views could be presented, but for now the reader may choose his or her favorite possibility, or none at all.

6. *Rainfall*

This chapter examines rainfall and associated phenomena and their possible relationship to solar activity. Rainfall can be measured directly using rain gauges or estimated by monitoring lake levels and river flows. Satellite and radar rainfall measurements have become increasingly important. Historical documentation on drought, or the absence of rain, also reveals empirical relationships. Both rainfall and evaporation show marked variations with latitude and geography. First, we examine these rainfall-associated variations and estimate how they might change with solar activity. Second, we cover empirical studies of rainfall, lake levels, river flows, and droughts.

A Theoretical Background for Rainfall Changes

The sun bathes the Earth's equator with enormous amounts of surface energy. Much of this absorbed radiant energy evaporates water, causes atmospheric convection, and is later released to space as thermal radiation. Steady-state energy escapes, so tropical temperatures do not rise without limit. Some absorbed energy is transported poleward by winds from the point of absorption. Intense convection near the equator leads to a large updraft known as the intratropical convergence zone (ITCZ), a band of lofty, high-precipitation clouds producing the largest rainfall of any region on Earth.

Solar energy in the ITCZ is carried to high elevations where it diverges and moves poleward. It is unable to travel all the way to the poles, so instead creates a large atmospheric circulation cell known as the Hadley cell. The Hadley cell has an upward motion near the equator and downward motions at about 30° north and south latitude. These downflow regions produce clear air with few clouds and create areas of minimum rainfall called deserts. These regions of upflow and downflow are connected by poleward flows in the upper atmosphere and equatorward flows in the lower atmosphere, forming a complete circulation pattern.

Outside the Hadley cell are temperate and polar regions. The temperate regions have more rainfall than the deserts, while the cold polar regions have even less precipitation. Figure 6.1 shows the three regions with relative maximum rainfall.

The mean evaporation has a much simpler latitudinal variation that tends to follow the surface temperature. Figure 6.1 shows this variation as a parabolic-shaped dotted line. the sun is about 0.15% warmer at sunspot maxima than at sunspot minima. Putting more energy into the Earth at sunspot maximum (1) causes the Hadley cell to expand and (2) intensifies the intratropical convergence zone, thus increasing the precipitation and lowering the mean surface pressure. One consequence of this is that the desert regions will tend to move poleward. The 0.15% additional radiation should produce about 0.15% more precipitation, or 0.15 cm/year, because the world average precipitation is about 100 cm/year. This increase would not be evident everywhere, but could be expected to follow a distribution pattern somewhat like that shown in Figure 6.2, in which the Hadley cell is allowed to expand so that the global increase in precipitation equals 0.15 cm. Most regions have increased precipitation, but desert regions may become drier. In all cases, the perturbations are so small their detection would be difficult. We now turn to rainfall records.

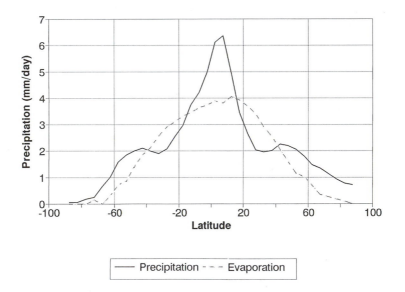

FIGURE 6.1 The mean latitudinal distribution of precipitation *(solid line)*, showing the high level of equatorial rainfall, and evaporation *(dotted line)*. Different authors give assorted values for precipitation and evaporation, so these curves are only approximate. However, all authors show similarly shaped curves.

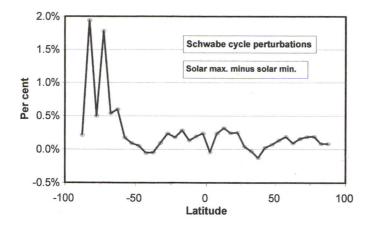

FIGURE 6.2 A plot of the percentage of perturbations in precipitation expected for a solar-cycle irradiance increase of 0.15%. We assume the Hadley cell expands. Most regions show increased precipitation, but deserts may become drier. In all cases, detection of these small perturbations is difficult. This simple model probably underestimates the effects near the equator where the intratropical convergence zone (ITCZ) intensifies. The Antarctic is a high-latitude desert region with almost no precipitation. The slightest perturbation can lead to large percentage changes, but the absolute changes remain small. Mathematically, this Antarctic effect is analogous to the changes in the tail of a Gaussian distribution discussed in chapter 4.

Rainfall

In the 1870s, more than 20 papers appeared relating sunspot activity to rainfall. These studies primarily concerned India, but considered any region for which data were collected. Locations investigated included Bombay and all of India, 54 locations in Great Britain, and 34 stations in the United States. All these early studies concluded that:

- When sunspots are numerous, rainfall is more plentiful.
- When sunspots are sparse, rainfall is sparser.

These early conclusions are consistent with theoretical expectations that the sun is warmer at sunspot maximum. In 1878, Chambers examined surface pressure statistics and rainfall records and concluded that the sun is hottest about the time when the spots are at a maximum.

With even the simplest theoretical models, an 11-year variation in rainfall should be evident in some regions but absent or with a reversed sign in others. Most regions will probably have increased rainfall with increased solar irradiance and increased sunspot numbers. This effect becomes clearer when examining pressure variations rather than simply looking at precipitation. Storms and high precipitation are associated with low air pressure; dry, clear weather is associated with high pressure. Because the total atmosphere has a constant mass, there must exist some regions where the average air pressure is higher at sunspot maximum. Other regions, however, must have lower pressures. H. H. Clayton (1943) states that the mean pressure is about 0.6 millibar (mb) less at the equator and about 0.3 mb higher near latitudes 50° north and south at sunspot maximum. This is fairly consistent with the data in Figure 6.2. Between 30° and 40°, Clayton finds almost no change, and above 65° he has no information. His picture of pressure changes would be consistent with our picture of precipitation changes if the 11-year solar cycle irradiance variations were larger than 0.15%, perhaps about 0.25%. Clayton's results are shown in Figure 6.3. In 1981 Gage and Reid showed that at the solar maximum the equatorial tropopause increased in height; that is, as more heat enters, the convection zone thickens. Surface pressure should decrease correspondingly, consistent with Clayton's findings. Although Clayton's findings are consistent with a solar forcing, the statistical significance of his results is unknown.

Summarizing the results to 1923, Clayton said that at the solar maximum, excess precipitation exists in the equatorial regions and a deficit in the midlatitudes. Excess precipitation is also indicated toward the polar regions. Figure 6.4 shows the geographical distribution of precipitation that is consistent with the simple theoretical picture we described, with a more intense Hadley cell and a stronger ITCZ. In Africa, the South Atlantic, and parts of South America the measured changes amount to more than 50 cm/year. This equals between 10% and 20% of average rainfall, which is higher than the few percent variations we expect. Perhaps the ITCZ is more intense, or perhaps local factors play predominant roles. An even larger percentage increase occurs near Iceland, perhaps resulting from shifts in the position of the Icelandic low, a semipermanent atmospheric feature. The sign and location of the precipitation

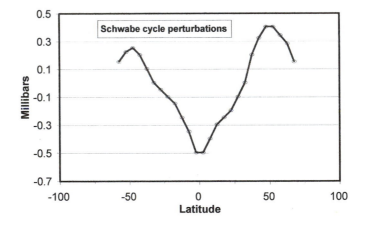

FIGURE 6.3 The changes in surface pressure at the solar maximum compared to the solar minimum (adapted from Clayton, 1943). In this figure, the net global pressure is unchanged and decreases in the equatorial regions where the intratropical convergence zone intensifies because of the brighter sun. Increased pressure in the mid-latitude regions balances the equatorial deficit.

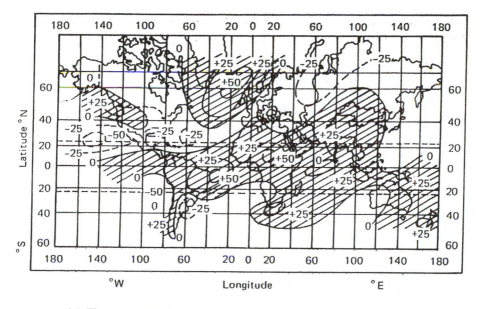

FIGURE 6.4 The geographical distribution of precipitation changes, comparing the sunspot maximum to the sunspot minimum (adapted from Clayton, 1923). Tropical regions *(dark)* have increased precipitation with increased solar activity and increased solar irradiance. Decreased precipitation occurs in some mid-latitude regions *(white)*, but polar regions may have increased precipitation.

changes are consistent with an increased solar irradiance at sunspot maximum. The magnitudes of the changes seem larger than expected. For the years 1860 to 1917, the precipitation variations are also inconsistent with the simultaneous temperature variations.

The scientific literature abounds with disagreements about what is happening concerning rainfall. For example, in 1878 Meldrum claimed Edinburgh, Scotland, exhibited a pronounced 11-year rainfall cycle. Yet when D. Brunt examined the same data in 1925, he found a 13-year cycle, not an 11-year cycle. Both Gerety and co-workers in 1977 and Deshara and Cehak in 1970 failed to find convincing evidence of a solar-cycle signal in precipitation records. On the other hand, in 1974 Wood and Lovett found that from 1540 to 1974 drought years in Ethiopia preferentially occurred at or near sunspot minima, a result consistent with theoretical expectations.

In 1994 Perry not only looked at statistical correlations but also provided a theoretical explanation. For simultaneous sunspot irradiance (see Figure 3.12) and precipitation variations, Perry finds a negative correlation over nearly all the continental United States. This correlation contradicts the positive correlation given by Clayton in Figure 6.4. Perry's highest correlation for precipitation occurs 4 years after the solar variations and are mostly positive values, particularly in the Pacific Northwest. He attributes the delay to a warming of the western tropical Pacific waters transported eastward over several years, changing the upwind boundary conditions off the coast of North America.

Some contradictions and inconsistencies among different authors can be resolved by viewing Figure 6.4 not as a static pattern but as one that undergoes slow secular changes over many decades. For example, the Icelandic low may gradually move east and west. Areas near the border of the positive-correlation region, such as England, may have positive correlations for a time, followed by negative correlations. The northeastern United States is also on the border between positive and negative correlations, so depending on the interval selected for an analysis, a positive, a negative, or even a null correlation may be found. This hypothesis is analogous to the positive and negative temperature correlation regions reported by Currie.

Just as he has done with temperature records, Currie has actively looked for cyclic variations in precipitation records. Figure 6.5 gives his power spectrum for precipitation for 1,203 locations in the contiguous United States. A weak 10.6-year peak is evident, but not nearly as strong as the 18.5-year lunar tidal cycle. After having examined many locations around the world, Currie generally finds the same effect everywhere: solar influences on precipitation appear to be less important than lunar tidal influences. Currie also uncovers many correlation sign changes as a function of location.

In summary, while earlier authors found a positive sun/precipitation correlation, many recent authors find no such correlation. Using the best analysis techniques, Currie found a small solar signal in precipitation for most locations. On the other hand, Perry is correct to consider time delays due to the thermal inertia of the oceans. Rapid responses are expected over land; a much slower response is expected for the oceans because of the higher heat capacity of the water. These delayed responses can influence the climate over land, particularly in coastal areas. Rain gauge observations for continental regions may produce a mix of direct and indirect effects. With the expansion and contraction of the Hadley cell, some regions will respond positively to a brighter sun while other regions simultaneously respond negatively. Geographical effects such as mountain ranges create further complications. Rain gauges produce immense problems when they systematically underestimate precipitation, have poor geographical distribution, and are redesigned numerous times over the years. All these problems make, finding an expected small sun/precipitation effect very difficult. If a real effect is occurring, for the most part the precipitation changes exhibit the correct sign and location. Yet different authors uncover such sufficiently contradictory results that a complete understanding of the problem still eludes the scientific community. Some sampling problems can be avoided by studying natural rain gauges such as lakes and river basins.

Theoretical Background for Lake Level Variations

Rainfall is quite variable spatially. One place may have an intense downpour, while a location only a few miles away has no rain at all. Because of this spa-

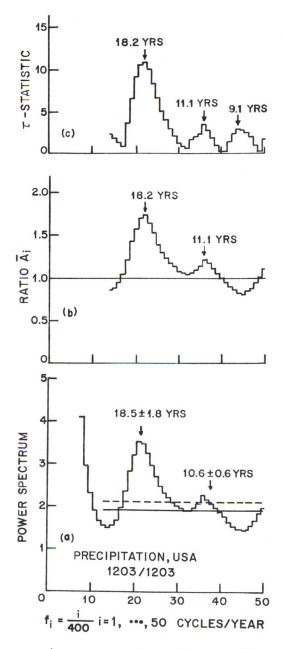

FIGURE 6.5 The composite power spectra for precipitation at 1,203 locations in the contiguous United States (*lower panel*). The weak solar signal is small compared with the lunar tidal signal at 18.5 years. The Saros cycle is 18.6 years. The upper two panels provide information on the statistical significance of the spectral peaks. (From Currie, 1992, with permission.)

tial variability, rain gauges do not always provide reliable regional means. Accurate results would require a very dense network of stations and, in the past, such networks were rare. Lakes, however, help us avoid this problem. As a first approximation, lakes can be treated as large rain gauges and their levels can be used to monitor rainfall. Still, in practice, complications arise. First, lakes lose water in a variety of ways, such as evaporation and stream run-off. For example, if a lake's level increases, its run-off may increase even more rapidly if the exit stream bed widens, allowing more water to be carried away. This could limit a lake's level to some maximum value and make the relationship between rainfall and lake level highly nonlinear. Additional complications may arise over many years as the outflow stream silts up and alters its flow or the river becomes dammed. Diverting lake waters for irrigation, diversions of incoming stream waters, changes in lake surface area and its concomitant rate of evaporation, and similar changes may all affect lake levels. Each lake needs to be considered as a separate entity to determine if a single physical mechanism controls its level. If a single mechanism is found, this mechanism may relate back to solar influences. Despite these many drawbacks, lakes are considered irresistible and are used to monitor rainfall. Our next section outlines the more than 15 studies that have been undertaken for African lakes.

Lake Victoria and Other Central African Lakes: A Classic Sun/Climate Relationship?

One classic tale in sun/climate relationships concerns Lake Victoria. As early as 1901, E. G. Ravenstein pointed out that the level of Lake Nyasa (or Nyanza) in Africa parallels the level of solar activity. In 1923 C. E. P. Brooks made a classic study of Lake Victoria and Lake Albert north of Lake Nyasa near the equator. Brooks's study showed a very strong correlation between the levels of these two lakes and solar activity from 1896 through 1922 (Figure 6.6). This .87 correlation implies that as much as 75% of the year-to-year variations may be solar-induced. Using anecdotal evidence, Dixey in 1924 indicated that the 11-year cycle for Lake Nyasa's levels extended from at least 1830 to 1923. Lake Victoria's level does not appear to be related closely to the measured values of precipitation, as shown in Figure 6.6, so Brooks argued that evaporation controlled the lake level. Presumably at solar minimum, when the lake levels are low, the sun would be warm. While this view is consistent with Koppen's temperature variations, it is inconsistent with the solar-irradiance variations seen in the last solar cycle.

G. T. Walker again examined the correlation between solar activity and lake levels in 1936. Walker argues that evaporation cannot be the controlling factor for these lakes because the local temperature and pressure variations are incompatible with such a contention, and he concludes: "It is interesting to note that on the whole since 1923 the levels of the African lakes have not varied in accord with the sunspot numbers." From this example some scientists have concluded that as soon as a sun/climate correlation is found and published, after

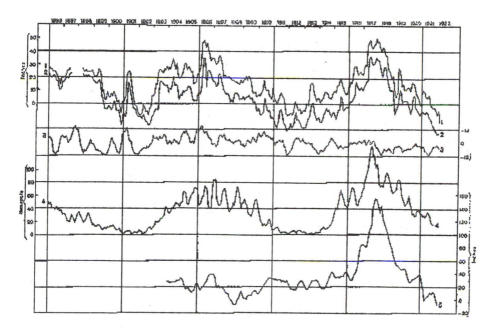

FIGURE 6.6 The parallel changes in African lake levels and solar activity as reported by Brooks in 1923. The upper two curves show the high and low Lake Victoria levels for 1896–1922. The next curve shows Ugandan rainfall. The fourth curve is the Wolf Sunspot Numbers. The lower curve is the level for Lake Albert. (Adapted from Brooks, 1923.)

that date it ceases to exist. By 1940 Clayton was pointing out that this correlation had reversed sign. In a 1983 discussion of this sun/climate "connection," J. A. Eddy updated the Lake Victoria level and sunspot number curves through 1972 (see Figure 6.7). Here the curves parallel each other only during the early years; recently, however, they are quite different. Although C. E. Vincent and his colleagues in 1979 reexamined Lake Victoria's level to find only that the correlation was now insignificant, they also found that nearby Lake Naviasha has a weak but significant correlation of −.32 with the Wolf Sunspot Numbers. Are we simply dealing with coincidences, or did a real sun/climate relationship disappear, perhaps because of some change in the sun? The previous chapter revealed that a large number of sun/climate relationships either broke down or reversed sign around 1920. Lake Victoria is a classic example of this effect. Many correlations simultaneously ceased all over Earth, suggesting global changes.

As mentioned, Dixey claimed that Lake Nyasa's level followed the solar cycle from 1830 to 1923. Although confirmatory data are lacking, after 1923 this correlation likely broke down. The Upper Shire River, which formerly drained the lake, gradually dried up, so by 1910 not even small craft could navigate it, and soon afterward the lake dried completely. The river bed also

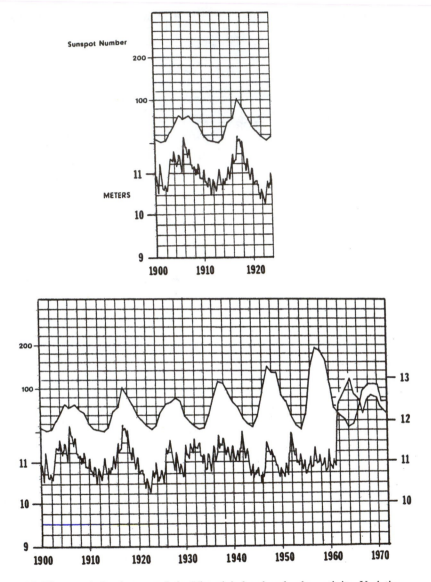

FIGURE 6.7 The correlation between Lake Victoria's level and solar activity. Updating the curves for Lake Victoria and sunspot numbers to 1972 reveals no evident relationship after 1923. The earlier correlation appears to be an accident. (Figure from John A. Eddy, with permission.)

silted up, so by 1924 rainwater flowed into rather than out of the lake. Lacking drainage, Lake Nyasa's level rose during the following years. Such effects probably destroyed any solar correlation, if one existed. These local changes reemphasize that each lake must be treated as a unique puzzle.

Australian Lake Levels

In 1923, C. E. P. Brooks examined the level of Lake George, a landlocked lake, with no outlet in New South Wales, Australia. Lake George's levels were measured from 1817 to 1919. For many consecutive years the lake can be dry. The lake achieved its maximum extent in the early 1820's, after which it shrank and dried up by the mid-1840's. It gradually recovered and regained its maximum extent, equal to its size in about 1823, in 1875. This last maximum was followed by a gradual drying, so the lake bed was again empty by 1902. However, by 1919 Lake George appeared to be refilling again. Brooks concluded that evaporation was probably the main controlling mechanism for the lake's level. Only about 15% of its variations could be attributed to sunspot numbers. According to Brooks, the rate of evaporation is greater when sunspots are few.

Most of the lake level changes appear to arise from the long-term secular variations described here. This implies that the weather was warm in the 1840s and from about 1900 to 1910, with cooler spells around 1823 and 1875. Lake George and Northern Hemisphere temperature variations correlated quite well during these years. Perhaps the variations in Lake George's levels reflect global changes that are not well measured by other techniques. In 1890 Bruckner pointed out that Lake Zurich, Lake Hamun-Sumpf in Persia, and the Great Salt Lake have major maxima and minima concurrently with Lake George. If solar variations controlled all these lake levels, this suggests high solar irradiances around 1845 and 1910 and low values around 1823 and 1875. Because these lakes are widely separately and show similar temporal behavior, an extraterrestrial effect may play a significant role. We will return to this point in chapter 10.

American Lake Level Variations

Many studies have been undertaken of U.S. lake levels and their possible relationship to solar activity. Perhaps the first such study was by G. M. Dawson in 1874. Dawson said high Lake Erie coincided with high solar activity. Before a search for a solar activity influence should commence, Nassau and Koski in 1933 imposed three criteria: (1) sufficient drainage area, (2) long-term measurements, and (3) adequate data on diversions by dams, locks, and power developments. Lake Erie, Lake Superior, Lake Ontario, and the Great Salt Lake all met these criteria. For Lake Erie, Nassau and Koski concluded that "two years after each sunspot maximum we have either a high or low lake level" and "where there is a variation in the period of sunspots there is a corresponding variation in the period of lake levels."

For nearby Lake Michigan, B. H. Wilson concluded in 1946 that from 1860 to 1944 the maximum yearly values have a distinct 11-year cycle. Wilson was apparently unaware of any other similar lake level studies, so his results have added significance. He corrected the variations in Lake Michigan levels for the water diversions caused by the Chicago Sanitary Canal, which opened in 1900, and the Sag Canal, which was completed in 1922. From the remaining

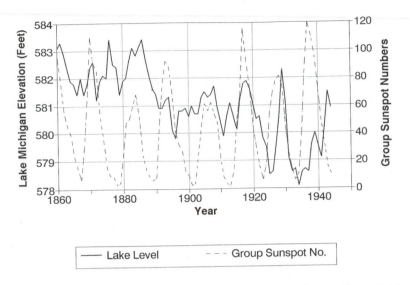

FIGURE 6.8 Lake Michigan elevations or levels for 1860–1942 (from Wilson, 1946). This lake's level is dominated by precipitation. According to Wilson, at times a distinct 11-year cycle is apparent.

time series, Wilson concludes that precipitation follows an 11-year cycle, with high precipitation associated with high solar activity. Wilson's results are plotted in Figure 6.8. On occasion, such as during the 1870s, the low correlation between solar activity and lake level seems to disappear.

In 1947 F. J. Ryder studied several Minnesota lakes. From meteorological records, Ryder concludes: "Years of large sunspot numbers coincide almost exactly with years of low temperatures at Minneapolis." In agreement with other studies, Ryder notes that Lake Traverse and Lake Superior are high at sunspot maxima. For Lake Superior, the correlation with solar activity is poorer than it is for Lake Michigan.

In summary, to a large extent precipitation governs the Great Lakes and nearby lakes. These same lake levels generally correlate positively with solar activity. This sun/climate relationship is weak and appears to have more academic than practical interest.

Theoretical Background for River Flow Changes

Like lakes, river basins may be viewed as large rain gauges. River flows provide an integrated measure of rainfall over the basin that the river drains. Monitoring river flows has certain advantages over using a network of rain gauges. First, the river acts like an amplifier, making small regional rainfall changes show up as large and easily measured changes in river flow. The river runoff comes from all basin regions and is eventually concentrated into a single channel that acts like an amplifier and integrator. For example, a 33% change in rainfall for the Devil Canyon watershed in California causes a 100% increase in run-off, an amplification factor of 3 according to Striett in 1929. River flows provide a single-point measurement, giving both large-scale averaging and amplification. These advantages may permit detection of a solar influence. Furthermore, these hydrological changes can have important practical applications for agricultural irrigation and hydroelectric power production.

Changes in the Nile River Flow

The Nile River empties a large portion of North Africa with water coming from as far south as Lake Victoria. The Nile's flow is largely controlled by the amount of precipitation in this large river basin. Monitoring the river's flow provides a measure of rainfall changes over a wide region. Commenting on earlier unnamed works in 1882, Balfour Stewart noted an 11-year cycle in the Nile River flow, as well as similar periodicities for the Elbe and Seine rivers. Between 1928 and 1987, Brooks, Streiff, Frolow, Riehl, Hassan, Hameed, Currie and others also made studies of the Nile River flow.

Using Fourier techniques, Hameed could not discern an 11-year signal in the Nile flow. If the signal does exist, it is weak. Using maximum entropy techniques, Currie finds cycles at 18–20 years and for 10–11 years for A.D. 680 to the present. Figure 6.9 shows his power spectrum for the summer floods for the years 1690 to 1962. Although the solar signal can be detected, it represents

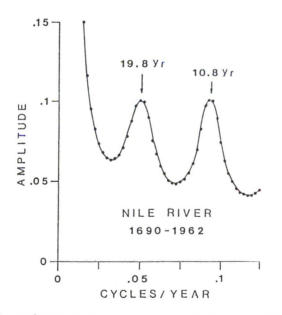

FIGURE 6.9 The maximum entropy power spectrum for the summer Nile floods from 1690 to 1962. As with many precipitation records, both a lunar and solar signal are present. Although the solar signal is clearly visible, it represents less than 1% of the observed variations. (From Currie, 1987, with permission.)

only a small portion of the observed variation. According to Hameed, less than 1% of the Nile variations would be explained by a 10- to 11-year solar cycle. Most variations are long-term and secular with long cycles, such as the one at 77 years.

Efforts to detect river-flow changes have engaged other scientists around the world. Between 1882 and 1992, Stewart, Lush, Shukow, and Sun and Yang conducted studies linking solar activity and river flows. These papers conclude that solar signals exist in river flows worldwide. Although these results could prove important economically, providing advance warning of increased probabilities of flooding, this field is not actively researched.

Drought Cycles

As early as 1889, D. E. Hutchins wrote a 136 page book called *Cycles of Drought and Good Seasons in South Africa* that claimed an 11-year solar cycle existed in South African droughts. These same droughts have recently undergone extensive study, and, while a 10- to 11-year drought cycle exists, the cycle is much weaker than the 18- to 19-year drought cycle. A 1987 examination by Currie of 194 drought/flood records from North America, China, and South America found 118 locations, or 61%, of the total had 10- to 11-year variations.

When discussing cyclic droughts, the most commonly mentioned cycles are 18–19 years or 20–22 years. For 288 worldwide drought/flood records, Currie (1987) found a 19-year cycle in 83% of the locations. Sometimes these drought cycles are associated with the Hale double sunspot cycle of 22 years and are attributed to solar changes. But Currie and others state that these drought cycles are caused by the 18.6-year lunar tidal variations. The classic case study concerns the so-called 22-year drought cycle in the western United States. Whether these droughts are solar-induced or lunar-induced phenomena is our next topic.

The Twenty-Two-Year Drought Cycle
in the Western United States

According to Hanzlik (1937), using only 30 years of data, the hydrologist Robert E. Horton in 1899 noted a 22-year cycle in run-off of Michigan's Kalamazoo River. This cycle is twice as long as the Schwabe cycle and is often referred to as the Hale double sunspot cycle because the sun's magnetic polarity has a 22-year cycle. Horton noted that the cycle continued in the following years and has been noticed in a variety of other phenomena, particularly droughts, in the western United States and Canada, but also shows up in storm tracks, precipitation, lake levels, and river flows.

Stanislaus Hanzlik of Prague wrote a short paper in 1937 called the "The Hale double solar-cycle rainfall in western Canada." He noted that this rainfall cycle is so strong, it will obscure any signal arising from an 11-year solar cycle and that the phenomenon deserved further study. Commenting on Hanzlik's paper, in 1943 Clayton said similar 22-year cycles were evident in Europe and Africa and thought they might arise from the alternate variations in the intensity of the solar cycle. Could the sun's brightness itself be alternately positively and then negatively correlated with solar activity? And is the 22-year cycle really 22 years rather than 18.6 years?

These questions were all brought into focus by J. Murray Mitchell et al.'s well-known 1979 study and several subsequent studies by others. Before turning to Mitchell's work, we would like to mention some precursor studies. In an important paper published in 1943, C. J. Kullmer measured the number of low-pressure systems crossing boxes of longitude and latitude in the United States. The western United States, and especially western Canada, displays a marked 20- to 22-year cycle in the number of storms. The earliest records, starting around 1880, show the most prominent variation. Assuming the variations are real, they imply a 20- to 22-year variation in precipitation and, hence, in drought arising from variations in the number and location of the storm tracks. As the amplitude of this variation has been damped out in recent years, it is now difficult, if not impossible, to discern. In 1945 C. J. Bollinger considered droughts in the Great Plains states and wrote: "The climate of Oklahoma and Kansas during the period of reliable meteorological record, 1886–1944, has exhibited a 22-year cycle of solar pattern. The recurrent series of wet and dry, good and poor crop years are not, as thought by some meteorologists, purely fortuitous, nor terrestrial, but mainly solar in origin, cyclic in character and hence roughly predictable." Baur, Krick, Leeper, Marshall, Miles and others also undertook studies touching on the Hale cycle and meteorological phenomena. Treating different phenomena and different intervals, most authors reported positive conclusions, while others reached negative conclusions concerning a 22-year climate cycle.

In the mid-1970s, Charles W. Stockton of the Laboratory for Tree Ring Research in Tucson mentioned to J. Murray Mitchell, a well-known climatologist with extensive research into climate change and its causes, that western U.S. tree rings may show a 22-year cycle of solar origin. If so, tree rings could be used to measure the spatial extent of drought regions from the 1600s to the present. Stockton and his colleagues had uniform-quality tree ring data spanning a long period and covering a large region. As a control set, these data were ideal for studying long-term climatic changes and for finding possible solar influences. Mitchell et al. (1979) related tree rings to the Palmer Drought Severity Index (PDSI), which provides a measure of soil moisture. The area of the drought is called the Drought Area Index (DAI). Tree rings could extend the measures of PDSI and DAI back to 1700.

Mitchell and his colleagues reached two main conclusions: (1) most of the time since 1700 the DAI expanded and contracted according to a 22-year cycle and (2) the severity of the droughts is proportional to the level of solar activity. These conclusions can be summarized by two figures from their 1979 article. Figure 6.10 shows a harmonic dial analysis of the DAI. In a harmonic dial, the passage of time is circular, completing a 360° circle every 22 years. For periodic events, the points will all line up along one radius vector. In this case, the

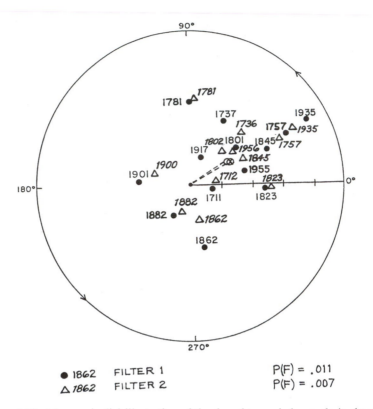

FIGURE 6.10 A harmonic dial illustration of the drought-area index cycle in the western United States (from Mitchell et al., 1979). Time passes cyclically and counterclockwise with a 22-year period. The drought peak dates are listed, using two different filters to find the maximum drought years. The further the maximum drought year is from the center, the greater the area of the drought. For example, the 1935 drought covered the most extensive region. (From Charles W. Stockton, with permission.)

points are the timing of the maximum extent of the DAI and its amplitude. The greater the amplitude, the further the point is from the center of the figure. Mitchell et al. used two smoothing filters on their data, one of 20.6 years and one of 24.3 years, bracketing the 22-year cycle. Both filters provide nearly identical results, indicating that the filter is not creating or significantly altering the cycles. Figure 6.10 reveals that most droughts tend to return every 22 years or so, suggesting a coherent cycle length lasting 360 years. Exceptions are the droughts of 1862, 1882, and 1901, all of which are out of phase with the others.

Mitchell et al. also concluded that the DAI is approximately proportional to the level of solar activity, as shown in Figure 6.11. Actually, the DAI precedes the envelope of solar activity. A better correlation can be found using an index like solar-cycle length. Despite this, Figure 6.11 suggests that the amplitude DAI is related to the sun, implying that the 22-year cycle is a solar, not a lunar, phenomenon.

More recent studies by Currie (1987) show that the 22-year drought cycle is actually 18.6 years in length and is caused by the 18.6-year lunar Saros cycle. This would explain why the droughts of 1862, 1882, and 1901 caused Mitchell difficulty. They are out of phase with the sun because DAI variations are not solar-controlled but lunar-controlled. Using maximum entropy methods, Currie finds an 18.6-year cycle in temperature for the eastern United States and western Canada. These cyclic variations are out of phase, indicating a standing wave pattern.

P. R. Bell argued in 1982 that a combination of solar and lunar effects caused the DAI variations. Bell also suggested that the real period of droughts is 20.5 years and is caused by a beat between the 22.279-year Hale double sunspot cycle and the 18.64-year lunar nodal tidal cycle. Physically, the lunar tidal cycle acts by controlling the area of surface waters that can be heated by the sun. These areas change with an 18.64-year cycle, leading to a 18.6-year cycle in the Earth's lower boundary conditions. In certain regions, such as the United States, these boundary condition changes cause climatic oscillations. The Hale double sunspot cycle is postulated to be the other forcing function. These two forcings beat together, sometimes enhancing their effects and sometimes canceling one another out. The net effect is a mean 20.5-year cycle, which still has considerable variation about its mean length. Bell predicts that

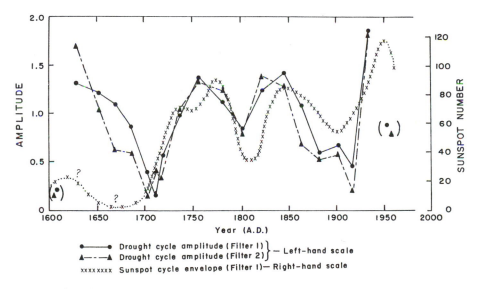

FIGURE 6.11 A plot of the smoothed drought-area index and the Wolf Sunspot Number (x's). The similar shape of both curves suggests a possible cause-and-effect relationship. However, the droughts precede the sunspot number curve, which is impossible if the Wolf Sunspot Number is the proper solar index to use for indicating solar influences on climate. A better correspondence between the sun and droughts would be found by using sunspot structure or sunspot cycle lengths, which are in phase with the drought-area index. (From Charles W. Stockton, with permission.)

the next Great Plains drought will achieve its maximum extent in 2005. Although Bell's ideas have appeal, the sun's irradiance is unlikely to have a 22.349-year cycle. With a 10.7-year solar-irradiance cycle, one expects beat periods at 6.8 years and 25.1 years. As an aside, Clough pointed out in 1920 that a 7-year cycle appears in various temperature, precipitation, pressure, crop yield, and other variables, particularly in the United States.

To date, there is no final explanation for the drought cycles in the Great Plains. Currie's results imply their timing is controlled by lunar tidal effects. Mitchell's results imply that the strength of the droughts is connected to solar variations. Bell argues that both lunar and solar effects contribute. The final answer awaits further study.

A Few Final Words

Until about 1923, measured variations in rainfall and associated variables followed a pattern spatially and temporally consistent with a warmer sun being a more active sun. The variations occurred in predictable locations, but their magnitudes exceeded theory and may be inconsistent with the global temperature variations that were lower when the sun was active. Alternatively, these so-called early global temperature variations may really be more representative of European temperature variations than global temperature variations. If the sun is always brighter when it is more active, then on average global conditions should produce warmer and wetter weather. A brighter sun would also be expected to create large regional variations with some regions being drier and some wetter. Since the 1920s, climate variations are consistent with a cyclic solar forcing of 10–11 years. The solar signal shows up in precipitation, lake levels, river flows, and droughts. In all cases, the signal is weak and does not explain much of the variation. Aside from academic interest, practical applications may include flood predictability, drought warnings, crop yield effects, and related matters. In many circumstances, even small advantages can prove beneficial.

Most precipitation and proxy precipitation records have much stronger cyclic variations, ranging from 18 to 22 years. This cycle implies that lunar tidal influences are more important than solar-irradiance variations in explaining most of the precipitation-related phenomena. To some extent, the sun appears to control the magnitude of the drought during the approximately 20- to 22-year drought cycle in the western United States, with both lunar tidal effects and the sun influencing the timing of these droughts.

7. *Storms*

We now examine some attempts to link storm numbers and storm track locations to solar activity. The number of both tropical cyclones and thunderstorms has increased and decreased with time and location as a function of solar activity. In fact, an early correlation between the number of Indian cyclones and solar activity proved so startling it caused an explosion of related research.

Modern Theories for Tropical Cyclones

In the previous century, tropical cyclones were called hurricanes or typhoons. Today tropical cyclones refer only to the weaker tropical storms with sustained winds above 31 miles per hour. Here, tropical cyclones refer to the stronger storms like those in the previous century. Anywhere from 1 to about 30 hurricane-strength storms can form each year. Among other factors, formation of these storms requires oceanic water temperatures above 26 °C (79 °F). William Gray at Colorado State University has successfully predicted the number of Atlantic Ocean hurricanes each year. This number is a function of the equatorial wind direction, the sea-level air pressure in the Caribbean, the strength of the westerly winds near the top of the lower troposphere, the presence or absence of an El Niño current, and, particularly, the amount of rainfall in the Sahel in Africa. Earlier we noted that increased solar activity produces a corresponding increase in rainfall in some regions. Figure 6.4 indicates that increased rainfall in the Sahel is expected, so based on this expectation and Gray's theory, hurricanes should increase in number. Higher solar activity and a higher solar irradiance can also be expected to increase the tropical ocean temperatures by a few tenths of a degree. These increased water temperatures tend to increase both the number of tropical cyclones and their intensity. Figure 7.1 illustrates the number of Atlantic Ocean hurricanes observed between 1962 and 1994 as a function of the sea-surface temperatures (SST). A sharp gradient exists in the

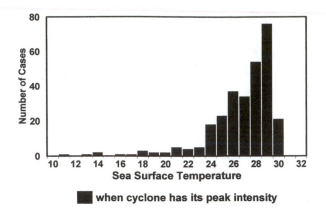

when cyclone has its peak intensity

FIGURE 7.1 The number of Atlantic Ocean hurricanes as a function of sea-surface temperatures. Increased temperatures lead to more and stronger tropical storms. Between 23 and 25 °C there is a sharp increase in the number of storms, suggesting that in some regions a brighter sun could lead to warmer sea-surface temperatures with more frequent and stronger tropical cyclones. (Data supplied by Mark DeMaria, personal communication.)

number of storms produced between 23 and 25 °C. In some regions, even a small increase in SST can lead to sharp increases in the number of tropical cyclones. Changes in solar brightness on the 11-year time scale could be expected to cause a corresponding cycle in the number and strength of tropical cyclones. Can this effect be detected, or is it damped in the random noise of the climate system?

The Number of Indian Cyclones: Meldrum's Cyclone Claims

In 1904 Sir Norman Lockyer commented:

> When I was preparing to go to India, in 1871, to observe the eclipse, Mr. Ferguson, the editor of the *Ceylon Observer,* who happened to be in London, informed me that everybody in Ceylon recognized a cycle of about thirteen years or so in the intensity of the monsoon—that the rainfall and cloudy weather were more intense every thirteen years or so. This, of course, set one interested in solar matters thinking, and I said to him: 'But are you sure the cycle recurs every thirteen years, are you sure it is not every eleven years?', adding, as my reason, that the sunspot period was one of eleven years or thereabouts, and that in the regular weather of the tropics, if anywhere, this should come out."

The 11-year cycle was subsequently confirmed, with 5 to 6 wet years followed by 5 to 6 dry years.

The above exchange may have been the inspiration for a large number of papers on the sun and climate that appeared in the following years. In 1872 Dr.

Meldrum wrote the first of a series of papers relating the number of tropical cyclones to solar activity. Figure 7.2, which is identical to Figure 1.1 in the introduction, shows the number of Indian cyclones each year and solar activity as represented by the Group Sunspot Number. The two curves have nearly identical shapes, with a correlation of .76, implying most year-to-year solar changes cause the variability in Indian cyclones.

The relationship shown in Figure 7.2 convinced many that the field of sun/climate relationships deserved serious consideration, and scientists began examining numerous other meteorological and climatic variables. In 1873 Poey found the same relationship for Caribbean hurricanes. In addition, the number of Indian Ocean shipwrecks was found to have an 11-year cycle. These findings led Blanford to state by 1891 that "among the best established variations in terrestrial meteorology which conforms to the sunspot cycle are those of tropical cyclones." Other scientists were less certain, and from 1870 on critics scoffed at the idea that the sun could have any effect on weather. Why these same critics were so adamant in dismissing sun/climate relationships appears uncertain, but most seemed convinced that the sun's radiant output was a constant (hence, the term "solar constant"). If the sun's output is indeed constant, they argued, it is easy to conclude that no solar mechanism could explain any of the observations. From the beginning of these studies, critics ruthlessly announced that no solar-induced explanation would be found. By the early 1900s, the critics seemed to have a point, as the strong correlation between the number of Indian cyclones and solar activity seemed to be disappearing (Figure 7.3). Around 1910, the number of cyclones peaked at a solar minimum—behavior opposite to that found when Meldrum first studied the problem.

A correlation that had existed for 70 years did not just cease to exist, but had changed sign. Recall that nearly every meteorological variable either

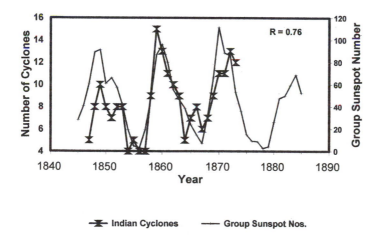

FIGURE 7.2 Meldrum's yearly number of Indian Ocean cyclones and the Group Sunspot Numbers. The two curves are very similar. This apparent correspondence has inspired many searches for similar relationships. (Adapted from Meldrum, 1872, 1885.)

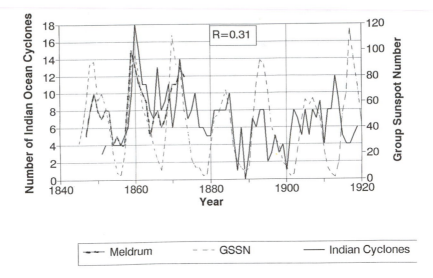

FIGURE 7.3 The number of Indian Ocean cyclones and Group Sunspot Numbers up-dated through 1920 using Visher's tabulations (1924). Some disagreement exists between Visher and Meldrum concerning the yearly counts, but the general trends in variations agree. By 1910 the positive correlation with solar activity has become negative.

changed its correlation sign or ceased to be correlated with solar activity during the period 1910 to 1930. The number of Indian cyclones is no exception to this rule. Cohen and Sweetser (1975) examined the variations in the number of Atlantic hurricanes as a function of time. Using maximum entropy spectral analysis, they found the power spectrum shown in Figure 7.4. A prominent peak in the number of cyclones occurs at 11.3 years. Cohen and Sweetser also found a peak in the number of cyclones at 51.3 years and a matching peak in solar activity at 52.7 years. The 133-year peak in the number of hurricanes is the maximum peak, but since the data record is only 100 years long, little confidence should be given to the length assigned to this peak. It may or may not correspond to the 95.8-year peak in solar activity that Cohen and Sweetser identify. These authors also find that the length of the hurricane season is related to solar activity, results that are consistent with a solar influence on climate, probably caused by increasing sea-surface temperatures due to an increase in solar radiation when the sun is active.

Paths of Mid-Latitude Storm Tracks

While discussing western droughts, we noted Kullmer's discovery that storm tracks crossing the United States change their locations during a 20-year cycle. Our reexamination of these same data finds that storm tracks were strongest in the western Canadian prairies during the 1880s, but have slowly damped out in recent years, becoming virtually undetectable.

In 1979 G. M. Brown and J. I. John examined storm track paths over the North Atlantic and Europe and found that their relationship to solar activity was a complex one. Basically, Brown and John found that the paths of northern mid-latitude cyclones changed position out of phase with solar activity. When solar activity is high, these tracks tend to follow a more southerly course (Figures 7.5 and 7.6). Over a typical solar cycle, the mean latitude of the storm track varies by about 1.5°. More southerly storm tracks, such as those crossing the Mediterranean, reveal the opposite relationship. When activity is high, the sun is brighter, so we expect weather systems to be pushed poleward. Mediterranean storm tracks follow this pattern. Why, then, do the northern storm tracks show opposite and seemingly paradoxical behavior? Brown and John note that when activity is high, the northern track splits into two tracks, with one track following the old path and a new track being created to the south. The average of these tracks causes the mean northern track position to move south. When solar activity is high, more northerly storms appear to exist, and these new storms are created in the region's south. This explains how, but not why, the paradoxical results in Figures 7.5 and 7.6 arise. Brown and John think that changes in the poleward temperature gradients and poleward transport of en-

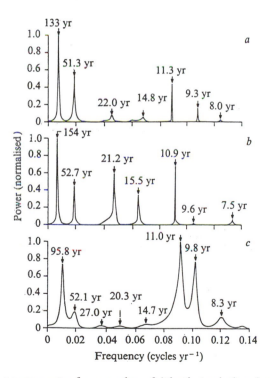

FIGURE 7.4 The power spectra for a number of Atlantic tropical cyclones, the length of the tropical cyclone season, and the Wolf Sunspot Numbers (from Cohen and Sweetser, 1975, with permission of *Nature*), suggesting that solar activity plays a role in the generation of Atlantic hurricanes.

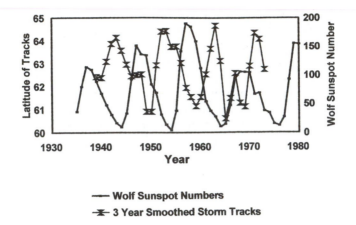

— Wolf Sunspot Numbers
✕ 3 Year Smoothed Storm Tracks

FIGURE 7.5 The average latitude of the storm tracks crossing longitude 15° W in the North Atlantic *(triangles)* is shown for 1938 to 1974, along with the Wolf Sunspot Numbers *(squares)* (data from Brown and John, 1979). The two time series are anticorrelated, and about 15% of the variations in yearly storm track positions might be explained by solar variations. A 3-year smoothing of the storm track positions makes the relationship easier to see. Brown analyzed only the winter storm tracks. For a typical solar cycle, the tracks change position by 1.5° between solar minimum and solar maximum.

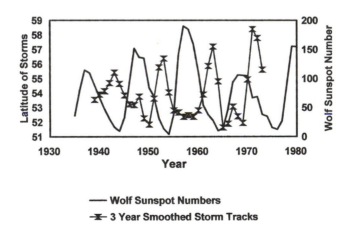

— Wolf Sunspot Numbers
✕ 3 Year Smoothed Storm Tracks

FIGURE 7.6 The average latitude of the storm tracks crossing longitude 5° E in the Baltic Sea *(triangles)* is shown for 1938 to 1974, along with the Wolf Sunspot Numbers *(squares)* (data from Brown and John, 1979). These two time series are also anticorrelated, and about 15% of the variations in yearly storm track positions might be explained by solar variations. A 3-year smoothing of the storm track positions makes the relationship easier to see. Brown analyzed only the winter storm tracks.

ergy are involved, but admit that the details of the process are unknown. The increased heat transport could create complex changes in the atmospheric circulation. Increased temperatures are often associated with more meridional circulation patterns, so the jet stream that controls the position of the cyclone tracks tends to bend north and south more markedly than at other times. These same meridional patterns could force storm tracks south in some regions and north in others.

This complexity has additional consequences. Depending on location, either more or fewer storm tracks may pass when solar activity is high. A lower temperature is expected with more tracks. Therefore, temperatures at these locations may be anticorrelated with solar activity. Other regions will demonstrate the opposite effect. This puzzling feature of sun/climate studies was pointed out in chapter 5 (see Figure 5.10). Creating regional averages might reveal a null effect if the correlated and anticorrelated subregions have equal areas. Areal averaging is not always advantageous when searching for sun/climate relationships.

Number of Thunderstorms

If the number and location of tropical cyclones and mid-latitude storms are related to solar activity, then could smaller storm features, such as thunderstorms, have similar relationships? Several controversial studies have reported an 11-year cycle in the number of thunderstorms. For example, in 1974 Stringfellow claimed the number of lightning incidences in England closely followed the sunspot numbers. He claimed that 64% of the year-to-year variations in lightning strikes are caused by changes in solar activity. Previously, in 1934, however, Brooks reported virtually no relationship between these two parameters in England, despite advocating such relationships. Brooks found less than 1% of the English thunderstorm variance could be explained by solar effects. More recent examinations of Stringfellow's dramatic results seem to suggest they are spurious, but that does not eliminate a less striking connection.

Australian Barry Pittock is a modern-day skeptic concerning sun/climate relationships, but for thunderstorms he says, "an association between thunderstorm frequency and the sunspot cycle is an inherently more attractive idea than many other proposed meteorological effects, since solar activity unquestionably affects low-energy cosmic radiation and the ionosphere. A physical link with thunderstorms via the atmospheric electric potential gradient seems but a small step in logic." Until now we have considered only one mechanism for sun/climate relationships—changes in the sun's brightness. Variations in the global electric circuit might very well be an additional contributing factor. Multiple methods may well exist by which solar changes can cause terrestrial changes. A more thorough discussion of these mechanisms is contained in J. R. Herman and R. A. Goldberg's 1978 book *Sun, Weather and Climate.*

Multiple possibilities exist for a connection between thunderstorm or lightning frequency and the sun. By becoming brighter, the sun can increase convection in the Earth's atmosphere with correspondingly increased numbers of thunderstorms and numbers of lightning strikes. This mechanism is compatible with a positive correlation between thunderstorm frequency and solar activity, but may not be sufficient from the energy standpoint. Some other mechanism must be present, such as cosmic rays, which are modulated by the sun's magnetic field. In 1979 Mae Lethbridge found that maximum thunderstorm activity coincided with maximum galactic cosmic radiation. Perhaps cosmic rays create ionization paths in the atmosphere, allowing more lightning to form along these paths. More thunderstorms would occur at these times simply because the Earth's atmosphere is more ionized by cosmic rays. The actual mechanism is unknown and involves many complicating factors. Cosmic rays also affect the ionosphere, perhaps modulating the conductivity of the high atmosphere. This would decrease the electrical resistance, allowing more current to flow not only higher up, but also in the troposphere. This higher current flow would be conducive to more thunderstorms. During 1978 Herman and Goldberg developed a theory that when cosmic ray activity is high, frequent solar proton events may cause increased ionization of the atmosphere above 20 kilometers. In certain regions, this enhanced conductivity could lead to increased current flows and an increased frequency of thunderstorms. Nonetheless, Ralph Markson pointed out in 1971 that it is not clear if changes in ionospheric electrical conductivity are causing the number of thunderstorms or whether the number of thunderstorms would cause the flow of current in the electrical circuit. For this problem, cause and effect are difficult to separate. Still, Herman and Goldberg suggest experiments to resolve this problem.

Since C.E.P. Brooks first introduced the idea in 1934, scientists have both confirmed and questioned the relationship between thunderstorm frequency and solar activity. Brooks tabulated the correlations of thunderstorms with solar activity for 22 regions around the world. Of these 22 areas, 5 had significant positive correlations. The best correlation occurred in Siberia, where 77% of the variance in thunderstorm numbers could be attributed to solar activity changes. Later studies, however, have not verified the Siberian results. Despite the contradictions and problems with the data, sufficient physical reasons exist to argue that the sun may exercise some control over thunderstorm frequency. The entire topic deserves more study.

Some Comments on Nonradiative
Sun/Climate Relationships

The global electric current plays a role in controlling the number of thunderstorms, so let us consider electrical and other nonradiative influences on climate. The November 18, 1882, edition of the *New York Tribune* contained the following quotation: "At the Mutual Union office the manager said 'Our wires are all running, but very slowly. There is often an intermission of from one to five minutes between the words of a sentence. The electric storm is general as far as our wires are concerned.' . . . The cable messages were also delayed, in some cases as much as an hour. The telephone service was practically useless during the day." That same day, the telegraph between Baltimore and Washing-

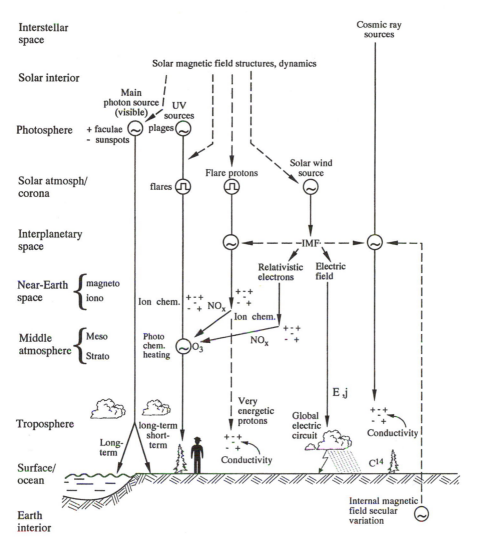

FIGURE 7.7 A diagram of the many mechanisms that may connect solar variations to changes in the Earth's weather and climate (from Roederer, 1995, with permission). However, we concentrate only on changes in the solar radiant output and its effect on climate.

ton worked using only terrestrial earth currents, without any need for batteries. Several times during the afternoon, Chicago's telephone switchboard caught fire. All these events occurred after a solar flare. Natural electrical flows can induce severe changes in electrical circuits. While we are primarily concerned with sun/climate relationships caused by changes in the sun's brightness, the above quotation provides a dramatic example of another sun-Earth connection. Considerable scientific literature describes the numerous nonradiative methods by which the sun could influence weather and climate. These complicated mechanisms may be called the Wilcox effect, the Brown effect, the Roberts effect, and so forth after their proponents, and may appear and disappear depending on other factors in the Earth-atmosphere system. Figure 7.7 summarizes many means by which the sun could influence Earth's weather and climate. These sun-Earth interconnections may be as important to changes in weather as those induced by changes in the sun's brightness, but will not be considered further here.

Some Final Words

The number of Earth's storms may increase with a corresponding increase in the sun's radiation. Theoretically, these increases may be subtle and difficult to detect, but empirical results reveal areas and intervals with very high correlations. The largest organized storms are tropical cyclones; thunderstorms are the smallest. Different physical processes may well control the frequencies of the storm's occurrences. Path changes for mid-latitude storm tracks would also be a natural consequence of solar climate forcing. Since the atmospheric circulation may simply be responding to the expansion and contraction of the Hadley cell, perhaps it is not surprising that the number and location of storm systems may have detectable 11-year cycles.

8. *BIOTA*

We now consider insect populations, circumpolar mammal populations, sea-weed density, agricultural yields, and similar topics. Good reasons exist to link such biological phenomena to solar activity. For one thing, if such meteorological parameters as temperature and precipitation vary with solar activity, life forms sensitive to small changes in these parameters may show dramatic responses. We will examine various claims from the 100 to 200 articles that either provide support for or criticize these types of ideas. The topics generally start at the lower levels of the food chain (i.e., insects) and proceed to the upper levels (i.e., predatory mammals), concluding with agricultural and economic studies.

Insect Populations—A Sensitive Climate Monitor

Insect populations are sensitive climate indicators. Paleontologists have used fossilized insects (see, for example, Coope, 1977) to show that very rapid changes in climate can occur in only a few years. Certain species of insects can tolerate only narrow ranges of temperature or precipitation. If meteorological variables alter that range, a new species of insect will replace the old. Insects occupy one of the lower rungs of the food chain, so fluctuations in their numbers may cause corresponding fluctuations in such predators as birds or spiders. Therefore, correlating insect populations with solar activity is a worthwhile venture.

In his doctoral treatise, "Über die Beziehungen der Sonnenfleckenperiode zu meteorologischen Erscheinungen" published in 1877, F. G. Hahn argues that locusts will probably appear in temperate regions only during unusually hot and dry years. Hahn shows that European locusts appear preferentially between the years of sunspot minimums up to the next sunspot maximum, an average of about 4 years. For the 7 years from the sunspot maximum to the next sunspot minimum, locusts are scarcer. Since sunspot minimums produced relatively warm temperatures for the years 1800–1862, this suggests that the sun influences European locust populations. E. D. Archibald, who in his later years was a very ardent advocate of sun/climate relationships, extended Hahn's findings. In a letter to *Nature* in 1878, Archibald showed that locusts appeared in Europe in 1613, 1690, and 1748–1749. According to Wolf, these dates occur 1 to 3 years after a sunspot minimum, which is consistent with Hahn's findings. In 1932 Criddle also reported that grasshoppers in Manitoba follow an 11-year cycle.

Independently of Hahn, A. H. Swinton in 1882 examined the number of butterflies captured each year in England over the course of four solar cycles. For 10 different species of butterflies, Swinton found that their maximum number occurred 2 years before the sunspot maximum, while their minimum number occurred just 2 years before the sunspot minimum. The numbers are quite noisy, but these results are nonetheless consistent with the locust population studies mentioned above.

Capturing butterflies to measure populations has it drawbacks because observers must maintain a constant interest and rate of capture. If their interest waxes and wanes, so, too, may the number of captures, even where butterfly populations are constant. Insect populations fluctuate markedly even in the reduced seasonal fluctuations of the tropics. Varying food supplies may cause population fluctuations, or the butterflies may be migrating in search of nourishment. Variations in both collector interest and fluctuating food supplies may mask or make difficult the detection of any cyclic behavior. From 1913 to 1941 New Jersey State Entomologist Thomas J. Headlee examined the density of eastern tent caterpillar nests, an easily recognized and rather immobile entity whose advantage is that changes in collector interest become less of a problem. Figure 8.1 plots Headlee's results along with the Group Sunspot Numbers. As in the previous studies, Headlee finds a peak abundance during the rising phase of the solar cycle. All these studies appear to have been made independently of each other and, although they involve different species, all have similar population histories.

In 1940 D. Stewart MacLagen made what was perhaps the most comprehensive study of insect populations and solar activity. Using accounts in newspapers, agricultural tracts, and almanacs of outbreaks of mosquitoes, antler moths, diamond back moths, leather-jackets, flea-beetles, and cutworms in Britain from the 1600s to the 1900s, MacLagen found an 11 year cycle. Insect population explosions tend to occur a few years before sunspot maximum, as shown in Figure 8.2. MacLagen speculates that these expansions in insect populations are caused by two factors: (1) increased warmth and precipitation and (2) increased ultraviolet radiation. Warmth and rainfall are conducive to an increase in insect populations, and increased ultraviolet light sparks greater activity in many life forms. This may be a kind of a bullwhip effect in which very small meteorological changes are amplified in the ecology, with insect populations being at the tip of the whip, so to speak. Birds and mammals, which are further up the food chain, would also be expected to have increas-

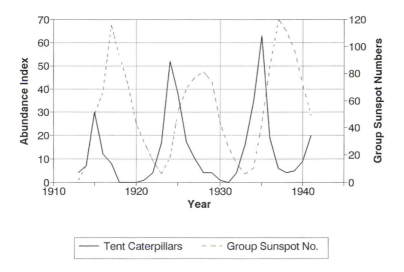

FIGURE 8.1 An index of tent caterpillar populations in New Jersey, measured by Headlee and plotted along with sunspot numbers. As with other insects, an 11-year cycle is apparent, but it is 2 years ahead of the solar activity. The out-of-phase relationship could be caused by some time constant in the biological system, or it could be that the sun tends to become brighter before solar activity peaks, such as is seen in the 1979–80 satellite measurements. (Adapted from Clayton, 1943.)

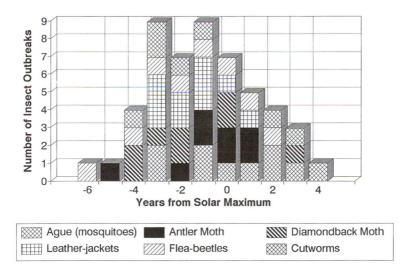

FIGURE 8.2 The frequency of British insect population outbreaks as a function of time relative to the sunspot maximum. Six British species are used (data from MacLagen, 1940). Note that the population outbreaks occur before sunspot maximum, just like the tent caterpillars in New Jersey.

155

ingly damped responses to solar variations. We now examine bird and mammal studies.

Bird Populations

Many birds feed on insects, so fluctuating insect populations may cause a corresponding change in bird populations. One problem arising from avian studies is that birds migrate, so at any one location their populations fluctuate due to both movement and intrinsic variations. Birds congregate where food supplies are abundant, but also respond to temperature and perhaps other meteorological parameters. The interaction could be quite complex and difficult to unravel.

In 1923 DeLury examined bird migrations in relation to solar activity. He looked at the arrival dates of cuckoos, larks, and swallows in Montdidier, France, from 1784 to 1869. DeLury found that at sunspot maxima, cuckoos arrived 9 days later than they did during sunspot minima years. These results are consistent with Koppen's study of temperatures for these years and suggest that when it is cooler at the sunspot maximum, birds migrate later. For larks and swallows, the spreads were 3 days and 1 day, respectively. Reexamining these data in 1936, MacLulich eliminated 24 years of data for cuckoos because they were incomplete and the spread was 3.75 days rather 9 days. A calculation of the correlation coefficient revealed no significant connection between sunspot activity and cuckoo and lark arrivals in Montdidier.

Apparently, there are no similar bird studies. Although not statistically significant, DeLury's results are plausible and what one would expect. It would be interesting to see if migration dates, or total populations, also fluctuate in a cyclic manner.

Circumpolar Mammal Populations

The *New York Times* of August 14, 1931, contained the following account of the Canadian Biological Conference held in July:

> In the plains around Edmonton, according to Dr. Rowan, a cycle of almost ten years is evident in grouse, other migratory birds and rabbits, and also in their enemies, such as coyote, lynx, red fox, and other fur bearers. Further north the voluminous records of the Hudson Bay Company had given Mr. Elton abundant data which showed a cycle of 9.7 years in hares, muskrats, grouse, lynx, red fox, marten, wolf, mink, and goshawks. Thus once in ten years or a little less, something seems to happen which causes an increase and then a decrease in the vital activities of both animals and plants all over North America from the borders of Alaska to the Maritime Provinces and northern United States, and also in adjacent seas. To complete the picture, Aurel Comisia, a Rumanian graduate of the Schemnitz Forestry School of Hungary, presented evidence of a ten-year cycle of disease in the rabbits of Europe.

It is now well known that during some years mammals such as rabbits and their predators are abundant, but are scarce in other years. Consider lynxes.

Lynxes feed primarily on rabbits. When rabbits abound, lynxes flourish and increase their population. When rabbit populations decrease, lynx populations collapse. This predator-prey pattern of expansion and contraction follows an approximate 10-year cycle. Ten years is sufficiently close to 11 years to suggest a connection to solar activity. DeLury first suggested this connection in 1923, which has remained controversial ever since.

Fortunately, a considerable body of apparently reliable data exists to study possible predator-prey abundances. Canada's Hudson Bay Company has for many years tabulated the number of fur pelts from lynxes and numerous other mammals. These records start in 1750 and continue to the present. For now we will assume that the number of fur pelts is proportional to animal abundances. We will look only at lynx fur counts, since they are more reliable than other counts, such as those of rabbits. Figure 8.3 plots the Group Sunspot Numbers versus lynx fur counts for 1820 to 1957. Keith's (1963) lynx count data for all of Canada are used after 1919, and data for the MacKenzie River valley, scaled to match the Canadian total, are used for earlier years. Data obtained before 1820 are considered less reliable and are not plotted. If we consider only the first three cycles after 1820, the correlation between the two plots appears good. Nevertheless, by 1855 the two curves are 180° out of phase. In fact, the two curves drift in and out of alignment, precisely the behavior one would expect from two unrelated variables having similar periods. The lynx fur cycle is about 9.7 to 9.9 years, compared with the 11-year sunspot cycle. From 1820 to 1957, 14 lynx fur cycles are plotted along with 12 sunspot cycles. From 1751 to

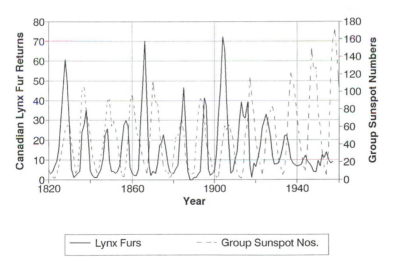

FIGURE 8.3 A plot of the Group Sunspot Numbers and Canadian lynx fur returns from 1820 to 1957 (lynx data from Keith, 1963). The two curves drift in and out of phase because they have different cycle lengths. Data after 1918 are from the MacKenzie River valley, scaled to match the Canadian totals.

1925, MacLulich in 1936 says, there are 18 lynx cycles and 15.5 sunspot cycles. From these observations, we conclude that no connection exists between lynx and rabbit populations and sunspot numbers.

Are the curves reliable? The sunspot values, and particularly their timing, are certainly reliable. The lynx fur counts for any given year are actually those lynxes harvested from the previous year. The fur count is then shifted by at least 1 year and sometimes 2 years for remoter regions from the actual population. Some regions adjusted for this shift, while other regions did not. These small accounting differences may account for some nonsimultaneity in different Arctic regions. Supply and demand for furs may also have an effect. For example, in 1892 Poland says that "Indians . . . trap all sorts of fur-bearing animals, and refuse to do business with a collector if he will not buy all the kinds."At times, the Indians keep their furs for a few years so that when the catch of new furs decreases and prices increase, they can sell. This activity suggests possible distortions, so the fur count may not always be a perfect representation of the population count.

A 1993 study by Sinclair and his associates at the University of British Columbia has new approach to estimating mammal populations. The lynx population depends on the rabbit population. Rabbits feed on the shoots of tree saplings. As rabbit populations expand, they eat more and more shoots and eventually consume their entire food supply. Most rabbits now starve, and numerous predators eat the rest. The rabbit and its dependent lynx populations both collapse, allowing the tree shoots to resume growing, starting the cycle anew. Sinclair and his associates use this model to trace rabbit populations in the Arctic during the last two centuries. They find that as rabbits nibble off the ends of the tree shoots, they leave scars that are preserved in the tree rings. By counting the density of these scars, scientists can deduce the population of shoot-eating snowshoe hares. Since 1750 the hare population closely parallels the sunspot cycle, especially when the sunspot cycle is strong. For the weaker sunspot cycles, the snowshoe hare cycle becomes closer to 10 years and diverges from the sunspot cycle. When high activity returns, the two cycles mesh together again. The linkage may derive from an alteration in weather. High solar activity may produce somewhat warmer and wetter weather, which is conductive to tree growth. Perhaps, on these occasions, this extra stimulation is sufficient to control the hare population through tree growth, locking solar activity, hare populations, and lynx populations in phase. Of course, more study is required.

Seaweed and Fish Variations

If land-based insect and animal populations are connected with solar activity, a similar relationship should be found for oceanic species. In 1956 F. T Walker found an apparent relationship between seaweeds off the coast of Scotland and solar activity for the years 1946 to 1955. Laminariacea are seaweeds that grow in the upper five fathoms of the ocean. The weight per unit area was measured

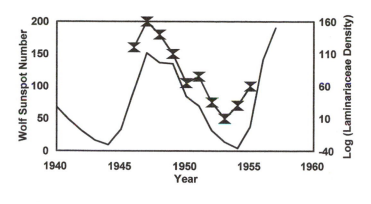

FIGURE 8.4 The mass density of Laminariaceae off Scotland and sunspot numbers for 10 years. The seaweed density seems to follow solar activity closely, but the data record is too short to be conclusive. (Adapted from Walker, 1956.)

both by in situ sampling and by areal surveys over approximately 1,000 square miles of water. Figure 8.4 reveals what might be a relationship between seaweed and solar activity but, to our knowledge, there has been no follow-up study.

Far fewer studies exist for ocean species than for land animal populations. As early as 1880, Ljungman claimed there was a 11.1-year cycle for the herring catch off Sweden. He also added that there was no apparent relationship between his curve for fisheries revenue and sunspot numbers.

Russian scientists have examined this problem in considerable detail. According to Izhevshii in 1964, the cod catches per hour of trawling by Canadians near Newfoundland show an inverse relationship to solar activity, with greater catches near sunspot minima. In 1973 Druzhinin and Khamyaova reviewed a number of Russian fishery records and claimed a relationship existed between fish catch and solar activity. Their arguments are difficult to follow and not convincing. This whole problem is complicated by a wide variety of other factors. Finding long-term data sets of fish catches is very difficult. Even if a brighter sun warmed the water, the fish would tend to migrate to new, more suitable regions. Thus, if the fleets followed the fish, no change in total catch would necessarily be found. To succeed in this type of study requires uniform catch records at fixed locations, with proper corrections for changes in technology or other effects. Since such an effort may not have been made, the correlation between fish catches and solar activity remains an open question. In any case, Walker's work suggests that follow-up studies would be worthwhile. Recently, in 1994 Wyatt and his colleagues found no solar cycle signal when examining Norwegian cod catches. Earlier, in 1993, Currie et al. discovered a weak 11–12-year cycle in European fish records. These subtle signals are not obvious in the raw data, but can be identified only by using sophisticated time-series analyses. Finally, the variations in seaweed are not in phase with the variations in lynx fur sales.

Tree Rings

Tree rings vary in width each year depending on variations in local temperature and local precipitation. Some tree ring growths are mostly controlled by temperature and others by precipitation, but generally the growth control a mixture of the two. Each tree must be studied to determine what factors are causing the variations in tree ring width. The astronomer A. E. Douglass began studying Arizona tree ring widths in 1901. In 1909 he claimed that tree rings exhibited cyclic variations of 11.3, 21.2, and 32.8 years, and attributed the 11.3-year cycle to sunspots. By 1919 Douglass had examined 75,000 tree rings from 230 specimens collected in Arizona, California, and the Baltic region of Europe. In his book *Climatic Cycles and Tree Growth: A Study of the Annual Rings of Trees in Relation to Climate and Solar Activity,* Douglass showed that the strongest solar cycle appeared in the Baltic region trees whose growth was governed by precipitation. Using tree rings of Flagstaff pines, Douglass claimed he could trace the solar cycle back 160 years. Douglass also used California sequoias. By 1919 he concluded that the sunspot cycle was evident in tree rings since A.D. 1400, except between 1650 and 1720 when the signal disappeared. These findings came to the attention of E. W. Maunder, who wrote to Douglass and pointed out that the sun had no spots from 1650 to 1720. To Douglass, this letter confirmed that his research was yielding correct and useful answers.

Douglass's 1909 results came to the attention of Yale economist Ellsworth Huntington, who was to become an active advocate of the idea that solar variations played a major role in climatic change. Huntington began looking at California sequoia tree rings extending back 2,000 years. He too stated that an 11-year cycle existed in tree ring widths. Huntington eventually wrote several books on the sun/climate relationship.

In 1926 J. A. Harris published a paper criticizing Douglass's results, saying the correlation between tree rings and sunspot numbers was positive but weak. This paper was the first criticism of the 11-year cycle in tree rings. Douglass himself continued to gather data and study tree rings. He found that the sunspot cycle often disappeared and that other cycles predominated. Today most scientists accept that there is little evidence of 11-year cycles in tree rings. For example, the 22-year drought cycle in the western United States advocated by Mitchell (chapter 6) was based on tree ring data. The power spectra for these tree ring data show little or no evidence for a solar-cycle signal. A more detailed study of tree rings and solar activity was carried out by LaMarche and Fritts (1972) using five different statistical techniques. LaMarche and Fritts found no evidence for a consistent or significant relationship between tree rings and solar activity. Currie in 1991 reexamined tree rings in North America using

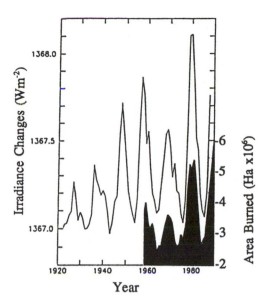

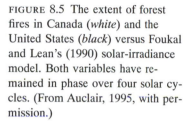

FIGURE 8.5 The extent of forest fires in Canada (*white*) and the United States (*black*) versus Foukal and Lean's (1990) solar-irradiance model. Both variables have remained in phase over four solar cycles. (From Auclair, 1995, with permission.)

his maximum entropy method and found a very weak 10.5-year cycle. Today we know that at many locations tree rings have small 11-year variations, but these cycles are not nearly as prominent as Douglass and Huntington believed.

Forest Fires

Increasing insect populations may attack and weaken certain trees. If a sufficient number of the trees die, this may exacerbate forest fires. Therefore, forest fires may have an 11-year cycle. Auclair (1992) examined this problem for Canada, Alaska, and the contiguous United States. His results, shown in Figure 8.5, reveal a clear 11-year cycle. Worldwide forest fire statistics have not been collected, but Auclair's work suggests a potentially more widespread relationship.

Wine Vintages

Wine vintages provide a good way to deduce the mean summer temperature at many European locations, as well as a homogeneous measure of climate going back to the 1400s. Early wine harvests suggest warm weather, and late harvests suggest cool weather. Several studies have been made concerning the relationship of wine vintages to solar activity. In 1979 Legrand and Simon found no relationship with solar activity. Although exceptionally early or late harvests tend to cluster around years of sunspot maxima and sunspot minima, this clustering is almost random. Nonetheless, other investigators disagree with Legrand's and Simon's conclusions, so the relationship between wine vintages and solar activity remains unresolved.

Agriculture and the Economy

Both agricultural yields and agricultural prices have been related to solar activity. Agricultural yields depend on precipitation and temperature, so a possible solar influence could manifest itself in crop yields. However, yields are also dependent on irrigation, fertilizer use, and insecticide application. In recent years, these modern technical developments have caused a partial uncoupling between yields and climate. Agricultural prices are a function not only of the yield but also of the area planted, government subsidies, the futures market, and global trading. So recent yield prices are even less likely to be influenced by the sun.

These technological advances mean that the best chance of finding a solar connection in climate will come from the earlier records. A number of positive results have been found, starting with Sir William Herschel in 1801. Herschel examined wheat prices and found that prices were higher when sunspots were scarce. He attributed the price increase to poor growing conditions caused by a cool sun. Herschel believed fewer sunspots produced a cooler sun because sunspots were hotter than their surroundings. Although his reasoning was faulty, his conclusion about the sun's brightness is consistent with recent satellite observations. In England, cooler weather may lead to poor crops and higher prices, but elsewhere this same reasoning does not apply. Warmer weather in certain regions can destroy crops through desiccation. No simple temperature-crop yield model applies everywhere. Since the wheat market was largely local before Herschel's time, his wheat price/solar activity connection still merits consideration. The most recent evidence, which involves a study of Beveridge's European wheat price listing for 1500 to 1869, shows no clear and dominant 11-year cycle. The most prominent peak occurs around 16 years. From this, we conclude that for Europe, at least, solar variations have played at best a minor agricultural role.

In 1927 C. C. Wylie found a negative correlation between Iowa corn yield and solar activity. L.P.V. Johnson found the same relationship for wheat, oats, and barley in Alberta and Saskatchewan, but the correlations were not significant. Coupled with Currie's work on temperatures, this suggests that high solar activity creates high temperatures that are detrimental to crop yields in western North America. In Scandinavia, Gloyne (1973) finds 11-year cycle in the length of the growing season, primarily controlled by variations in springtime temperatures. His results are compatible with a brighter sun when it is also an active sun. In 1988 Currie found a 10.0-year cycle in Iowa crop yields, a 9.5-year cycle in Arkansas, and a 10.9-year cycle in Illinois. These cycles are far weaker than the 18- to 20-year cycle caused by western droughts. Only the Illinois cycle seems possibly solar-induced. By chance, one expects to find some region exhibiting a cyclic variation close to the length of the solar cycle, which may be true here. Curiously though, and in support of real effect, both the crop yields and the mammalian populations in Iowa and Arkansas exhibit 9.5- to 10.0-year cycles.

All these diverse studies suggest that a solar-cycle signal may be present in crop yields and perhaps in crop prices. It is worthwhile examining the wider implications of the solar effect on agriculture. In the past, agriculture occupied a significantly larger economic role, so agricultural booms and busts could lead to prosperity or poverty for the entire economy. India demonstrates this fact in striking fashion. During the 1870s, India was an English colony. In an 1878 debate concerning India before the British House of Commons, Lyon Playfair argued, "it was established that famines in India came at periods when sunspots were not visible. Out of twenty-two great observatories of the world, it had been shown in eighteen that the minimum rainfall was at times when there were no spots on the Sun." Playfair based his comments on Meldrum's studies. If true, a change in rainfall caused by a change in solar activity could cause famines in India and cause both the Indian and British economies to suffer, as well.

These questions are certainly worth considering. In 1878 Sir Stanley Jevons, a famous British economist, began seeking connections between the economy and the sun. Between 1700 and 1878 Jevons identified 14 commercial crises in England that, because they were spaced about 10.5 years apart, could have arisen from solar variations. Jevons thought that a bright sun led to drought in India, thus decreasing demand for foreign manufactured goods and causing a commercial panic in England. Although plausible, 7 of the 14 years Jevons lists for crises occur close to solar minima, 4 close to solar maxima, and 3 not clearly associated with either a maxima or minima. In the earlier years, the commercial crises tended to cluster around solar minima: in the later years around solar maxima. Like the changes in sign between correlations of temperature and solar activity, this latest relationship shows similar long-term secular sign variations. By his own admission, Jevons received a fair amount of ridicule for studying only this problem. If a solar/commercial cycle exists, as Jevons claimed, it is not a straightforward one.

In 1934 Garcia-Mata and Shaffner considered Jevons's theory for economic cycles. They thought that with more data and with improved methods of examining time series they might find a basis for Jevons's hypothesis. Their study produced two surprises: first, Garcia-Mata and Shaffner found no evidence that the agricultural economy fluctuated in parallel with solar activity; second, as with solar activity, all of manufacturing and production fluctuated with an 11-year cycle. Figure 8.6 shows their results. This figure also suggests that recent commercial crises are again clustering around sunspot minima. Garcia-Mata and Shaffner could not explain this result, nor can we. No fundamental link seems to exist between solar activity and its influences on agriculture and manufacturing. Garcia-Mata and Shaffner suggest that changes in psychological factors, caused by changes in the global electric field, may lead to long-term shifts in optimism and pessimism that show up in the manufacturing economy. Their fundamental conclusion is that the entire problem needs more study, and people should not be quick to dismiss the correlations as coincidences.

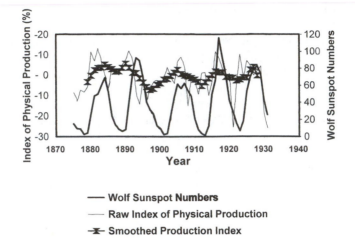

— Wolf Sunspot **Numbers**
— Raw Index of Physical Production
⭐ Smoothed Production Index

FIGURE 8.6 A redrawn plot of Garcia-Mata and Shaffner's index of physical production (%) versus solar activity (Wolf Sunspot Numbers). A 7-year running mean was used to create the smoothed index *(triangles)*. Garcia-Mata and Shaffner suggest that psychological factors, rather than physical factors, are producing the parallel changes.

Today, the United States Federal Reserve System continues to fund solar-economic studies, which indicate that the reasons for the correlation—whether accidental, psychological, or physical—remain to be found. Jevons's initial ideas may yet prove of some merit.

Final Comments

In this chapter, we have considered various proxy biological and related phenomena that could logically respond to changes in solar radiation through alterations in weather and climate. These include insect populations, bird migrations, seaweed density, fish catches, Arctic mammal populations and fur catches, tree rings, wine vintages, agricultural yields and prices, and the economy in general. In many instances, the conclusions are discouraging, but a sufficient number of positive findings should encourage further work. Most promising are those phenomena occurring at the base of the food chain, for example, insects on land and seaweed in oceans. Current data concerning these phenomena appear insufficient either to confirm or to refute a solar activity connection.

9. *Cyclomania*

In his 1874 book *Contributions to Solar Physics,* Sir Norman Lockyer writes the following:

> Surely in meteorology, as in astronomy, the thing to hunt down is a cycle, and if that is not to be found in the temperate zone, then go to frigid zones, or the torrid zones and look for it, and if found, then above all things, and in whatever manner, lay hold of, study it, record it, and see what it means. If there is no cycle, then despair for a time if you will, but yet plant firmly your science on a physical basis, as Dr. Balfour Stewart long ago suggested, before, to the infinite detriment of English science, he left the Meteorological Observatory at Kew; and having got such a basis as this, wait for results. In the absence of these methods, statements of what is happening to a blackened bulb in vacuo, or its companion exposed to the sky, is, for research purposes, work of the tenth order of importance.

As Lockyer notes, looking for cycles is certainly an attractive prospect. If found, a cycle will help with predictions, and successful predictions are a central goal of scientific studies. Cycle hunting is also a relatively straightforward procedure, with several well-developed techniques. Cycle hunting has become even easier with the advent of computers. By feeding a stream of data into an algorithm to detect cycles, one is likely to find cycles even in a series of random numbers. Hence the danger that a cycle, if detected, may prove to be only a random fluctuation. Will the cycle persist with more data, or will it simply disappear? For ardent cycle hunters, and the sun/climate field has attracted its fair share, these questions are hardly a deterrent. The lure seems to be finding something of seeming importance with minimal effort.

From a practical point of view, a cycle may be considered important only if it can be plotted. If sophisticated analyses are required to detect the cycle, the cycle probably has only secondary importance. While these criteria are not the usual mathematical criteria for significance, they are a practical, down-to-earth guide to what is important. Judged by these criteria, the solar sunspot cycle is highly significant because it is easily seen in the data. In contrast, some of the 11-year temperature cycles are discerned only after extensive analyses. Nevertheless, from the scientific standpoint, these difficult-to-detect cycles are still important because they may indicate a casual relationship.

When judging the usefulness of any prominent cycle, the paramount question is whether a physical mechanism exists to explain it. If not, the cycle may arise purely by chance. Furthermore, newly emerging ideas in chaos theory require one to recognize that "cyclical" phenomena in chaotic systems, such as the turbulent terrestrial and solar atmospheres, may not exhibit strict periodicities. Instead, near certain frequencies their power will not be strictly periodic. Thus, any search for a strict periodicity may be doomed to failure, but intelligent searches for power near certain frequencies may lead to valuable results. We now discuss certain cycles that might be expected to predominate in the sun/climate problem.

The 27-day Cycle

As viewed from the Earth, the sun rotates with a period of approximately 27 days. Occasionally, more large sunspots and active regions occur on one hemisphere of the sun than the other. This east-west asymmetry may persist for several months, modulating the Earth's solar irradiance with a 27-day period. This 27-day period seems to be more persistent at the shorter wavelengths in the solar spectrum because shorter wavelengths arise high in the solar atmosphere. Since the ultraviolet radiation reacts most strongly with Earth's upper atmosphere, most of the 27-day responses occur high in our stratosphere and mesosphere. The time constant for the Earth-atmosphere-ocean system is so long (4–5 years) that it damps out any short-term response in the atmosphere's lower levels. Since 1844 occasional claims of 27-day changes have been made concerning solar influences on the Earth's atmosphere. Twenty-seven day responses may be confused with the 29-day lunar orbital period that could affect meteorology through tidal forces. Because of the small expected response from solar forcing, its transient existence and potential confusion with lunar effects, searching for a significant solar-induced signal of 27 days in weather is probably not worthwhile.

The 154-day Cycle

Recent studies show that a small, 154-day cycle now exists in many solar phenomena. To date we know of no attempt to find a meteorological response. As with the 27-day cycle, a weak response is expected. Nineteenth-century reports suggested a 233-day solar cycle that now no longer seems apparent. Could cycles between 100 and 300 days arise by chance and exist for several years or several solar cycles?

The 10- to 11-year Schwabe Cycle

This obvious solar physics cycle has often been reported in meteorological phenomena. The meteorological cycles might, however, reverse sign or disappear. During some cycles sunspot blocking might dominate the solar radiative output, while during other cycles facular emission is the dominant component. If the dominance shifts back and forth, it could yield very confusing meteorological responses, an area that certainly deserves more study.

The 18.6-year Lunar Saros Cycle

This lunar tidal cycle shows up in a variety of meteorological records such as temperatures and, perhaps, droughts. The Saros cycle may affect climate by altering the amount of tidal water and its solar heating. In addition, this cycle may interact with other cycles to produce beat cycles. For example, the beat frequency between the Schwabe and Saros cycles could yield cycles at 6.8 to 6.9 years and at 25 to 26 years. Seven-year cycles have been noted in many meteorological phenomena.

The Hale Double Sunspot Cycle of 22 Years

Many meteorological parameters such as droughts and storm tracks seem to occur in a 20- to 25-year cycle. The sun's radiative output is unlikely to have a cycle of this length, so the cause of these meteorological cycles must be sought elsewhere. Some records clearly show an 18.6-year lunar nodal tidal cycle. Perhaps in short meteorological records an 18.6-year cycle is mistakenly being identified as a 22-year Hale cycle. Alternatively, perhaps changes in the sun's magnetic polarity are affecting the Earth's electrical system, which, in turn, influences certain aspects of the Earth's atmosphere.

The Gleissberg Cycle of 80 to 90 Years

This cycle seems fairly clear in the sunspot record and in its proxy measurements by cosmogenic isotopes. The cycle appears to show up in many meteorological parameters, suggesting that there may be an important sun/climate connection over long periods of time.

The 180- to 208-year Seuss Cycle

Carbon-14, which responds to solar variations, has a reported cycle of around 200 years. This same cycle also shows up in such other climate proxies as the oxygen-18/oxygen-16 ratio used to measure oceanic temperatures, and tree rings, which respond to precipitation and temperature. These results suggest a sun/climate relationship. Because long time scales are involved, self-consistent time series for both solar inputs and potential terrestrial responses are difficult to find. Even longer cycles, variously reported as 300 years, 400 years, 1,600 years, and so forth, have been hypothesized. The origin of cycles greater than or equal in length to the Gleissberg cycle (80–90 years) is not well understood. What feature in the sun could maintain its integrity for such long periods so that these cycles could persist in phase? What memory mechanism could the sun have? These two questions currently have no satisfactory answers. Perhaps these long-term features are more akin to persistencies than cycles.

Some Claimed Sun/Weather/Climate Cycles

In contrast to these one lunar and six solar cycles, scientists have attributed about 40 terrestrial cycles to solar influences. Several cycles are listed in Table 9.1. Some cycles may be solar-induced even though they are not detailed

TABLE 9.1 Table of some claimed solar-induced cycles in climate.

Cycle Length	Parameter(s)	Claimant
6.6456 days	Temperature	Abbot, 1947
6–7 days	Temperature	Clayton, 1894
7 days	Precipitation	Abbot, 1939
25 days	Temperature/pressure	Asakura, 1954
27 days	Many	Many
28 days	Halos	Archenhold, 1938
5 months	Temperature	Takahashi, 1954
9 months	Several	Clough, 1936
355 days	Cyclogenesis	Berkes, 1954
2.5 years	Many	Clough, 1936
3 years	Pacific cycle	Berlage, 1934
3.2 years	Climate	Goschl, 1928
5.25 years	Climate	de Boer, 1938
5.6 years	Climate	Polli, 1950
7 years	Pacific cycle	Berlage, 1934
8 years	Thunderstorm frequency	Aniol, 1952
11 years	Many	see Chapters 5 to 8
11–12 years	Many	see Chapters 5 to 8
13 years	Thunderstorm frequency	Aniol, 1952
18 years	Norwegian glaciers	Rekstad, 1924
20 years	Temperature	Willet, 1950
21 years	Thunderstorm frequency	Aniol, 1952
22 years	Many	see Chapter 7
22.75 years	Many	Several
23 years	Several	Several
46 years	Climate	Abbot, 1937
50 years	Varves	Bradley, 1929
80 years	Interdiurnal temperature	Berkes, 1955
89 years	Budapest climate	Thraen, 1949
92 years	Climate	Abbot, 1937
100 years	Climate	Memery, 1935
300 years	Climate	Clough, 1905
334 years	Climate	Auric, 1936

above. For example, the sunspot cycle is asymmetric, rising fast and slowly tapering off. Performing a Fourier analysis on these asymmetric cycles reveals such higher harmonics as 5.5 years and 2.75 years, which show up as weak periods simply because the use of symmetrical trigonometric functions like the sine and cosine requires the higher frequency variations to reconstruct the asymmetric sunspot curve. If the meteorological response has the same asymmetry, a Fourier analysis will also show up with cycles of one-half, one-quarter, and so forth of the basic 11-year cycle. Some cycles in the table may be responses to this type of effect. The 23- and 46-year cycles could be higher harmonics of the 80- to 90-year Gleissberg cycle.

As mentioned previously, other cycles could arise from the beat between two cycles of different lengths. Some 7- and 8-year cycles may arise from beats between two other cycles. The 2.5-year cycle in the table is known as the quasi-biennial oscillation (QBO). The QBO arises strictly from internal interactions between the Pacific Ocean waters and the atmospheric winds. No solar forcing is postulated for these phenomena.

Finally, satellite microwave measurements have revealed a 7-day cycle in the Earth's temperature. The cycle's amplitude is 0.02 °C, so it is small and difficult to detect. Today this cycle is attributed to mankind's thermal input operating on a 7-day schedule. As early as 1894, however, Clayton was reporting a 6- to 7-day temperature cycle that he attributed to solar influences.

We end by again cautioning: (1) the detection of a cycle does not prove that the cycle is based upon a physical mechanism; and (2) the sun/climate field has an enormous number of studies that may be classed as forms of cyclomania. Even a detected cycle with a period similar to a known solar period does not prove the existence of a sun/climate connection. The detection of cycles is important because it may provide an impetus to developing physical theories to explain them.

III. THE LONGER TERM SUN/CLIMATE CONNECTION

10. *Solar and Climate Changes*

Until now we have considered only 11-year variations in solar activity and climate. The sun also varies on longer time scales. Since these variations seem to parallel a number of climatic changes, the sun may contribute to climatic changes on time scales of decades to centuries. We now examine several solar indices that vary in parallel with Earth's climate change. There exist plausible arguments that these indices are proxy indicators of the sun's radiative output, but there is no proof. We now present the strongest correlations we have seen for a sun/climate connection. First, as it is the most widely publicized index, we consider the mean level of solar activity. In 1801 Herschel first proposed a relationship between climate and the level of solar activity. Second, we examine solar cycle lengths, which have been studied sporadically since 1905. Third, we look at two closely related indices—sunspot structure and sunspot decay rates. Fourth, we consider variations in the solar rotation rate. Lastly, we examine some major solar and climatic events of the last thousand years to see if any indications of solar influence are evident on climate. Although we present the solar-induced changes as arising from total-irradiance variations, as discussed earlier spectral-irradiance changes may be the primary driver.

Mean Level of Solar Activity

When Rudolf Wolf reconstructed solar activity based on historical observations of sunspots, he found an 11-year cycle going back to at least 1700. In 1853 Wolf also claimed that there is an 83-year sunspot cycle. This longer term variation becomes evident simply by smoothing the data, as in Socher's 1939 example (Figure 10.1). Wolf's original discovery of an 83-year cycle was forgotten, but the long cycle was rediscovered by H. H. Turner, W. Schmidt, H. H. Clayton, and probably others. After W. Gleissberg also discovered this 80- to 90-year cycle around 1938, he published so much material on the subject

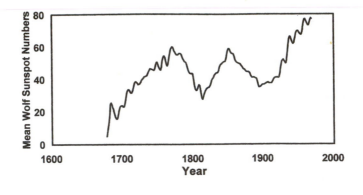

— Peaks in 1770 and 1851

FIGURE 10.1 A 45-year running mean of the Wolf Sunspot Numbers is plotted follow-ing Socher, who showed a similar plot in 1939. The so-called Gleissberg Cycle of 80–90 years is evident from peaks in 1770 and 1851. The most recent peak was about 1968, so it is not clear if the Gleissberg Cycle has a constant length.

that ever since it has been called the Gleissberg cycle. All these rediscoveries of the same phenomenon indicate that the 80- to 90-year cycle may be real but not strictly periodic. Rather, the cycle may be a "persistency" with an 80- to 90-year period. During this period solar activity is quite powerful but fails to exhibit a single sharp spectral peak.

From the extant historical sunspot records, only about three 80- to 90-year cycles can be discerned. These long cycles are too few to tell us much about the cycle or even if it is real. Recent techniques using cosmogenic isotopes have reconstructed solar activity on longer time scales. Two isotopes are com-monly used—carbon-14 and beryllium-10, both produced by cosmic rays. Ga-lactic cosmic rays are modulated by changes in the strength of the interplane-tary magnetic field arising from changes in solar activity. Hence, measurements of the isotope concentrations as a function of time provide proxy measures about the history of solar activity.

Beryllium-10 (^{10}Be) has three advantages over carbon-14 (^{14}C):

1. Beryllium-10's residence time in the atmosphere is 1–2 years ver-sus 20 years for ^{14}C. With less lag and smoothing, the solar activity variations can be discerned more clearly.
2. Beryllium-10 gives high time-resolution reconstructions.
3. In the twentieth century, Beryllium-10 is not as strongly influenced as ^{14}C by mankind's activities.

Because of these advantages, we will look at the ^{10}Be record first. Three figures illustrate the ^{10}Be results. Figure 10.2 shows the smoothed variations in the Wolf Sunspot Numbers and the ^{10}Be concentrations from Greenland's Dye 3 core. This plot reveals that the two curves generally agree from at least 1750 to 1900; after 1900 mankind's activities may influence the results. In particular, both quantities show the Dalton Minimum around 1800. This encouraging similarity offers hope that we can study solar activity during the centuries before telescopes existed.

Figure 10.3 shows both sunspot and ^{10}Be data after a spectral bandpass filter between 9 and 13.7 years in length was passed through both data sets. This procedure enables us to look at the 11-year signal in both data sets. First, consider the Wolf Sunspot Numbers. The outside envelope of the curve indicates the overall level of solar activity. The envelope is very narrow for the Dalton Minimum around 1800, indicating a low level of solar activity. The ^{10}Be data also shows the Dalton Minimum. Yet for the lesser solar activity minimum around 1900, the ^{10}Be data suggest a maximum in solar activity even larger than the great solar activity maximum of 1957–1958. In this case, the ^{10}Be data clearly provide an inaccurate proxy for the Wolf Sunspot Number. On the other hand, the ^{10}Be level for 1750–1800 indicates a lower level of solar activity than does the Wolf Sunspot Number. As mentioned in chapter 2, this result seems to occur when the Wolf Sunspot Numbers are too high. So for 1750–1800, the ^{10}Be data seem to provide a better indication of the solar activity level than does the Wolf Sunspot Number. Yet for 1750–1800, the ^{10}Be data show three solar cycles, while the sunspot numbers have four cycles. To us it seems that the ^{10}Be reconstruction missed a solar cycle. All these discrepancies become important when we relate them to climate. These relationships, if they exist, are sensitive to the amplitude and timing of the cycles. If solar activity cannot be reconstructed correctly, properly relating this activity to climatic changes also proves difficult. Each technique used to reproduce solar activity has some associated error, and these limitations and errors must be kept in mind.

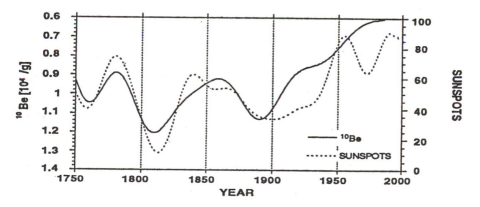

FIGURE 10.2 Long-term variations in the ^{10}Be and Wolf Sunspot Numbers. Low-pass filters of 41 years and 31 years were applied to the respective data sets. From 1750 to 1900, the two smoothed curves appear similar, but they diverge after 1900. A ^{10}Be response to processes not represented by sunspot numbers may cause this divergence. (From Beer et al., 1993, with permission.)

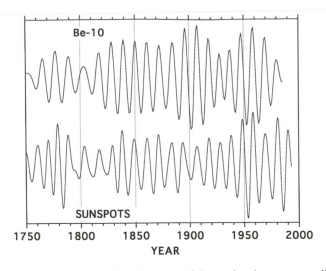

FIGURE 10.3 The Schwabe cycle band. A 9- to 13.7-year bandpass was applied to the ^{10}Be data and Wolf Sunspot Numbers. The outside envelope corresponds to the smoothed curves given in Figures 10.1 and 10.2. The Dalton Minimum around 1800 shows up in both records, but the sunspot minimum around 1900 does not appear in the ^{10}Be record. Also note that the ^{10}Be record has three full cycles from 1750 to 1800, compared with four full cycles in the sunspot records. (From Beer et al., 1993, with permission.)

The value of the ^{10}Be solar activity reconstruction lies in its potential to extend our knowledge of solar behavior back many centuries. Figure 10.4 shows the ^{10}Be filtered data from 1453 to date and presents additional puzzles. In chapter 2 we noted that very few sunspots were observed between 1650 and 1700. The telescopic observations provide little room for ambiguity. Yet the ^{10}Be data display a well-defined and strong 11-year cycle for this same period. Perhaps both data sets are correct, but are not measuring the same thing. Even without sunspots, an 11-year cycle may occur in other solar phenomena such as coronal extent and general magnetic field strength. Furthermore, these same field-strength variations could still strongly modulate the ^{10}Be production. In short, the ^{10}Be concentration seems to parallel solar activity as measured by the sunspot numbers only some of the time. When sunspots disappear, the ^{10}Be may or may not be a good proxy. The Spoerer Minimum around 1500 provides a reason to believe there were few sunspots. The ^{10}Be concentrations seem to confirm this Spoerer grand minimum, while failing to confirm the grand minimum known as the Maunder Minimum. Therefore, the ^{10}Be record must be used with caution during both solar and sun/climate studies.

In addition to the ^{10}Be record, we also have the ^{14}C record. How well do these two records of solar activity correspond? Figure 10.5 shows measured ^{14}C concentrations along with simulated ^{14}C concentrations, using the ^{10}Be data input to a carbon budget model by Siegenthaler and Oeschger. The two curves show reasonable agreement, except perhaps from 1650 to 1700 when the mea-

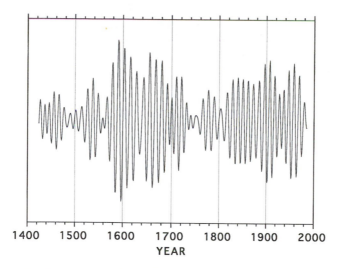

YEAR

FIGURE 10.4 The Schwabe cycle band. The 9- to 13.7-year bandpass filter was applied to the ^{10}Be data from 1423 to 1985, using the ^{10}Be data from the Dye 3 core in Greenland. Note the strong 11-year signal in the Maunder Minimum (1645–1715). No doubt few sunspots were observed then, so the ^{10}Be record must be responding to an aspect of solar activity other than visible sunspots. Low ^{10}Be amplitudes exist around 1480–1520, an interval known as the Spoerer Minimum that, like the Maunder Minimum, probably had few visible sunspots. (From Beer et al., 1993, with permission.)

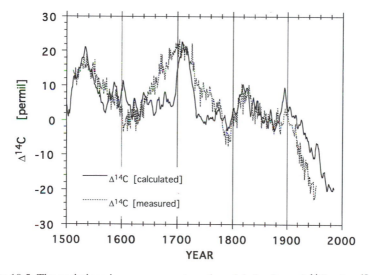

FIGURE 10.5 The variations in measurements and modeled values of ^{14}C using ^{10}Be input and a carbon cycle model by Siegenthaler and Oeschger (1987). Here, high values of ^{14}C correspond to low values of solar activity. Note the extremely high ^{14}C values around 1710. From 1690 to 1700 sunspots were seen on only 4 days—the deepest portion of the Maunder Minimum. The ^{14}C deviation probably corresponds to this same decade, but it is delayed 20 years due to the carbon system's slow response. (From Beer et al., 1993, with permission.)

177

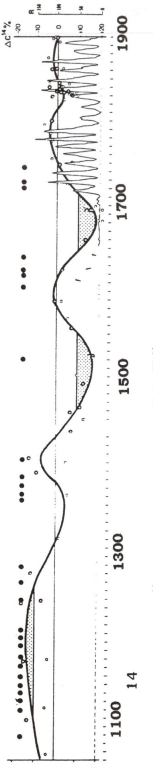

FIGURE 10.6 Measured values of ^{14}C (*solid curve*) plotted in parallel with the Wolf Sunspot Numbers. The Maunder Minimum (1645–1715) is clearly evident, but the Dalton Minimum (1795–1825) is less clear. The slow ^{14}C response time tends to obscure smaller variations such as the Dalton Minimum. Also note the Spoerer Minimum around 1500, another grand minimum around 1350, and a Medieval Maximum around 1200. Dark circles are naked eye sunspots from Kanda (1933). (From John A. Eddy, with permission.)

sured ^{14}C more clearly shows a lower level of solar activity. The higher values in Figure 10.5 correspond to lower solar activity. When activity is high, the extended solar magnetic field sweeps through interplanetary space, thereby more effectively shielding the Earth from cosmic rays and reducing the production of ^{14}C. Low solar activity lets more cosmic rays enter the Earth's atmosphere, producing more ^{14}C. In effect, Figure 10.5 displays an inverted mean level of solar activity. The ^{14}C data reveal that major solar events, such as the Maunder Minimum, are clearly evident. Figure 10.6 (from Eddy and Boornazian, 1979) shows inverted ^{14}C concentrations proportional to solar activity and plotted together with the Wolf Sunspot Numbers. The last thousand years reveal two, and possibly three, grand minima: (1) the Maunder Minimum from 1645 to 1715, (2) the Spoerer Minimum around 1500, and (3) an unnamed minimum around 1350. The Grand Maximum in solar activity near A.D. 1200 is also evident. Eddy and Boornazian provided these names in 1976 when he rediscovered the great paucity of sunspots in the late 1600s, and he named the Maunder Minimum in honor of E. Walter Maunder who publicized its existence for more than 30 years.

Before Eddy's work on this topic, the scientific community in general was unaware of how few sunspots existed in the late 1600s. This anomalous solar behavior is often mentioned in the literature, but for the most part the phenomenon was either ignored or dismissed as a measurement error. For example, in 1942 Luby quotes as follows from Annie Maunder's book *Splendor in the Heavens*: "The long dearth of sunspots lasted from 1645 to 1715, or 70 years, and during this time, so far as we know, there were no spots large enough to be seen by the naked eye; all required the use of a telescope." To this claim, Luby responds: "It is difficult on reading this and other summaries of the 'seventy year period of quiescence' to avoid the conclusion that here the trouble is dearth of observations and not dearth of spots." Perhaps Luby's sentiment was rather common, so the Maunder Minimum received little attention. Eddy's careful work changed this attitude. First, he revealed that many telescopic observations took place during the Maunder Minimum, and, second, he showed that the ^{14}C record provided another proof for the existence of a grand solar activity minimum.

As with any idea advanced in the scientific literature, one can ask: Why was this result published at this particular time? Often the answer is simple curiosity, but sometimes there are other reasons. In this case, Eddy claimed that the dearth of sunspots during the Maunder Minimum and the Little Ice Age were related. It appears that when the sun was quiescent for long periods, the Earth often cooled. When solar activity reached particular heights, such as the Medieval Maximum or during the twentieth century, the Earth often became warmer. Beyond its meteorological effects, the warmer Earth during the Medieval Maximum produced several social effects, such as Greenland's flourishing Viking colony. To Eddy, these parallel solar and terrestrial variations were far more than a coincidence. Eddy advanced the hypothesis that the envelope of solar activity provided a measure of the solar radiant output. Using this envelope of activity, or a good proxy for it such as the ^{14}C record, one could then

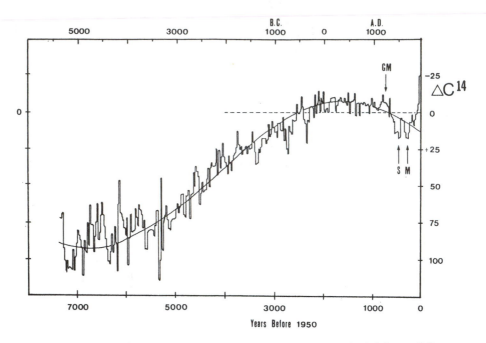

FIGURE 10.7 The ^{14}C record for the last 7,000 years. The Maunder Minimum (M), Spoerer Minimum (S), and Medieval Maximum (GM) are shown by letters and arrows. Over long time scales, the Earth's magnetic field modulates changes in the ^{14}C flux, as shown by the sinusoidal curve. Downward deviations from this sinusoidal curve correspond to extended solar minima, showing that, on average, they occur every few hundred years. (From John A. Eddy, with permission.)

extend the forcing-of-the-Earth back many thousands of years. Figure 10.7 extends the ^{14}C record back 7,000 years. Large downward excursions corresponding to grand maxima occur frequently. Each downward excursion should correspond to a cooler climate. The connection between these large solar excursions and climate helped motivate Eddy to publicize the Maunder Minimum. In the early 1970s, the prospect of global warming due to carbon dioxide renewed scientific interest in climatic change, and Eddy's work became very pertinent and attracted much interest. Many ideas similar to those in Eddy's 1976 work had been published as early as 1905, yet, like the Maunder Minimum, these early works were overlooked. These earlier works form the basis for our next subsection.

In the past two decades, additional evidence has arisen that both supports and weakens different aspects of Eddy's sun/climate concepts. Eddy presents the basic idea that the sun's radiant output parallels the envelope of activity. If this is so, the sun's output should be lower during the Maunder Minimum than it is now. The level of both solar and stellar activity can be measured by examining the calcium II K line that is sensitive to variations in magnetic activity. Researchers at Kitt Peak Observatory used a measure of line strength called the

K index and found that this index varies from about 0.17 Å to slightly more than 0.20 Å. Similar results came from measurements taken at the Sacramento Peak Observatory. However, for solar-like stars, whose ensemble is thought to represent the sun's variations if it could be observed for many centuries, the K index varies from about 0.13 to 0.21 Å (see Figure 10.8). These results appear to show that the sun's last cycle is about as active as the sun ever becomes. Like other stars exhibiting Maunder Minimum–like conditions, the sun's activity can become much lower than observed in recent decades. The secondary peak just below 0.1485 Å probably shows those stars that have entered their grand minima. That the K index and solar total irradiance parallel each other suggests that when the sun enters one of its grand minima, its irradiance shows a corresponding drop. With a 0.15% variation in total irradiance and a 0.03 Å variation in the K index over the last solar cycle (indicated by the solid boxes in Figure 10.8), the K index scale can be converted to a solar-irradiance scale. In 1992 Judith Lean and her colleagues made this calculation and found that during the Maunder Minimum the sun would have been 0.25% less bright than it was during the solar minima of 1985–1986. Such a variation is large enough to have some impact on climate and suggests that the late 1600s should have been cooler than the present. During the Maunder Minimum, the sun's disk would have been free both of faculae and an active network to achieve these low solar-irradiance values.

In 1993 Rind and Overpeck used the GISS GCM to model the consequences of a prolonged decrease in solar irradiance such as might occur during the Maunder Minimum. They find that while most of the world cools, certain regions are actually warmer (see Figure 10.9). This simulation reveals warmer regions east of Scandinavia or around the Black Sea. In general, European

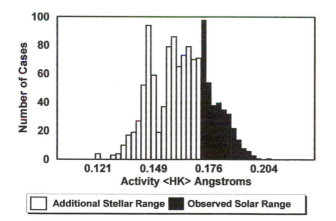

FIGURE 10.8 The solar Calcium <HK> II index or equivalent line widths for solar-like stars *(both open and solid bars)* and for the sun *(solid)*. Note that the sun has varied only over the top portion of the range represented by stars, suggesting that a longer examination of the sun would reveal larger variations in solar activity and probably irradiance as well. (Adapted from Baliunas and Jastrow, 1990.)

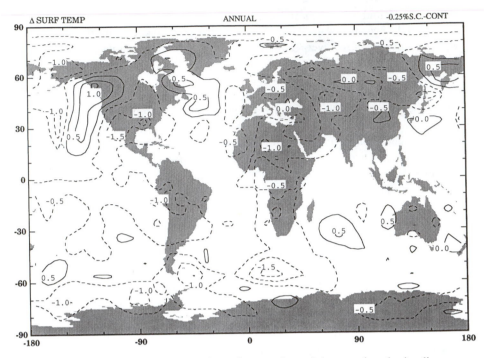

FIGURE 10.9 Temperature perturbations after a prolonged decrease in solar irradiance, such as might occur during a Maunder Minimum–type event, based on the GISS GCM (from Rind, 1995, with permission). Note that some regions may actually be warmer.

cooling is very weak, with perhaps the strongest effects in England and Spain. Yet one must proceed with caution when interpreting past climate records. For example, though tree ring records from a single Scandinavian site were examined and warmer conditions were found during the Maunder Minimum, this does not prove that the entire planet was warmer. Regional effects can obscure or even cancel global signals.

Anecdotal evidence from a particular region or location is often used to support the idea of a cool Maunder Minimum. Eddy pointed out that Dutch paintings during this period often portray snow in places where snow is seldom seen today. In the 1600s, the Thames River often froze, now a rare event. John Goad reports three freezings of the Thames in 1662, several freezings during the very cold winter of 1665–1666, more ice on the Thames in 1667, 10 days of ice in 1677, 9 days of ice in 1678, 2 months of freezing in 1678–1679, and, for the first time in human memory, the Thames froze below the city bridge in 1683. This anecdotal evidence concerning repeated freezings of the Thames suggests a cooler England during the Maunder Minimum, but does not provide proof because of the great variability that naturally occurs at any one location. These freezings are consistent with the GISS GCM results.

Groveman and Landsberg, as well as Bradley, have attempted to reconstruct past climates. In 1979 Groveman and Landsberg used all available temperature measurements to reconstruct and model the variations in temperature of the Northern Hemisphere from 1583 to the present. The further back they went in time, the more uncertain their results were since the measurements became fewer and were confined to Europe. Groveman and Landsberg also found that, compared with other times, the Maunder Minimum did not appear particularly cold, which in turn suggested that there is little evidence for solar forcing. A similar analysis by Bradley in 1992, using a variety of proxy climate indicators, found that the period before the Maunder Minimum was cooler than the Maunder Minimum itself, again suggesting no sun/climate relationship.

Another problem that arises when using the envelope of solar activity as a proxy for solar irradiance occurs when the Northern Hemisphere shows its peak warming in the 1930s while the solar activity envelope peaks in the 1950s. Figure 10.10 illustrates this discrepancy. The envelope of solar activity often appears to lag behind the variations in air temperatures, the opposite of what would be expected for a true relationship. This plot appears to suggest that using the envelope of solar activity as a proxy for the solar total irradiance variations is too simple. The failure of the solar activity envelope to track measured air temperatures is a major drawback to Eddy's hypothesis. Not everyone would agree with these conclusions. For example, in 1991 George Reid, pointed out that sea-surface temperatures and the envelope of solar activity track each other well (see Figure 10.11). However, recording of sea-surface temperatures is sporadic and based mainly on ship observations. Further, mea-

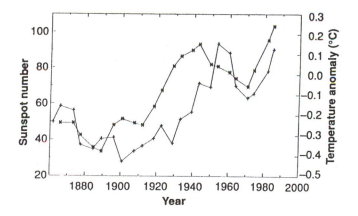

FIGURE 10.10 The smoothed Northern Hemisphere air temperature anomalies and the smoothed level of solar activity plotted together (from Friis-Christensen and Lassen, 1991, with permission). Air temperatures peak about 1940, but solar activity does not peak until some 15 years later. This plot tells us that cause and effect might be reversed, or that simply using the peak of solar activity as a proxy for solar-irradiance variations does not appear to be valid.

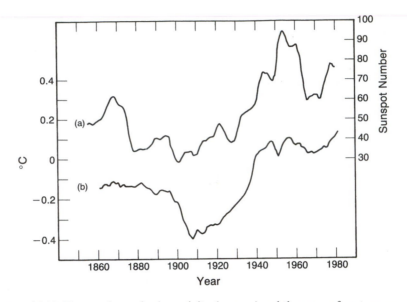

FIGURE 10.11 The envelope of solar activity *(curve a)* and the sea surface temperatures *(curve b)* track each other well (from Reid, 1991, with permission). Despite this
apparent success, the more accurately measured air-surface temperatures over land
change about 20 years before the corresponding change in the envelope of solar activity (see Figure 10.10).

surement techniques have changed dramatically over the years, making the construction of an internally self-consistent data set difficult. Perhaps the measurements are correct. The time response of the oceans may be sufficiently slow
that a long delay between solar forcing and sea-surface temperature response
occurs. Reid's correlation is certainly intriguing and requires further study.

Solar-Cycle Lengths

In 1890 the Swiss professor E. Bruckner published a lengthy monograph on a
35-year cycle in weather and climate. Using routine weather observations, records of glacial advances and retreats, the occurrence of severe winters, the
times of grape harvests, and so forth, Bruckner deduced that a regular 35-year
cycle existed in every parameter he studied. We call this the Bruckner cycle
and today, if it is known at all, it is not believed to be true. Recent tree ring
analyses in Scandinavia do have a prominent 35-year cycle that persists for
many hundreds of years, so this cycle may yet prove real. Bruckner could not
explain the cycle but thought it was probably induced by solar changes. What
solar variability index has a 35-year cycle?

As early as 1901, N. Lockyer and W. J. S. Lockyer suggested that changes
in solar-cycle lengths were an index that could be correlated with the Bruckner
cycle. Based on limited data, Lockyer noted that three short cycles tended to

alternate with three long cycles. H. W. Clough extensively investigated Lock-yer's suggestion, publishing between 10 and 20 thorough articles during his lifetime. Three of Clough's papers are important to us: (1) "Synchronous varia-tions in solar and terrestrial phenomena," in 1905, (2) "The 11-year sun-spot period, secular periods of solar activity, and synchronous variations in terrestrial phenomena" in 1933, and (3) "The long-period variations in the length of the 11-year solar period, and concurrent variations in terrestrial phenomena" in 1943. In these papers, Clough anticipated many modern developments in sun/climate relationships.

Clough's 1905 article first discusses 12 meteorological variables that con-tain Bruckner cycles. The second section discusses a 36-year cycle in solar phenomena, including variations in solar-cycle length. Using the Wolf Sunspot Numbers for the years 1600 onward and the Fritz compilation of cycle epochs based on auroral and Chinese records, Clough compiles a list of cycle lengths from A.D. 300 to 1900. From these records he deduces both the cycle-length variations and Bruckner cycle length variations. Clough further extends his re-sults to consider variations in the length of 83- and 300-year cycles. At this point Clough reaches two important conclusions:

> "The solar-spot activity is periodically accelerated and retarded, and this action is primarily manifest in the varying length of the 11-year spot cycle."

> "When the [sunspot] period is shorter, the sunspot number is larger."

Clough also notes that variations in cycle length and sunspot number are not perfectly in phase, but that variations in sunspot number lag behind the varia-tions in cycle length.

In 1933 Clough amended and corrected his 1905 paper, writing: "The rela-tive numbers vary inversely with the length of the 11-year period with a lag of about 10 years, while the ratios vary directly with a lag of about 15 years." To illustrate Clough's conclusions we replot a portion of a figure from his 1933 paper in Figure 10.12. Here Clough shows the variations in solar cycle lengths smoothed using seven sunspot periods. He smoothed his data to emphasize the existence of an 83-year cycle. Such smoothing is dangerous and can introduce nonexistent periodicities. Apparently, Clough's smoothing procedures did not alter his analysis. Clough's original figure showed the variations in the fre-quency of aurorae, the frequency of severe winters, the frequency of Chinese earthquakes, the rate of sequoia tree growth, and the rise and fall of the lower Nile River. He apparently intended to illustrate the existence of 83-, 87-, 78-, 80-, and 77-year cycles in these various phenomena. He would make similar arguments regarding the 37-year, or Bruckner, cycle and the 300-year cycle. Figure 10.12 shows the cycle length and frequency of severe winters contained in the original plot.

In our opinion, Clough's work overemphasizes the search for cycles. His articles contain so much material, they become overwhelming. Nonetheless, one basic 1943 conclusion was: "The efficient cause of weather variations seems to be some form of solar energy, one manifestation of which is a varying

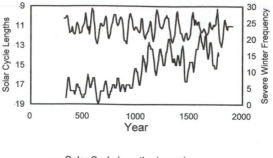

— Solar Cycle Lengths (upper)
— Frequency of Severe Winters (lower)

FIGURE 10.12 Variations in solar-cycle length and the frequency of severe winters as given by Clough (1933). Although Clough himself did not calculate any correlations, he claimed there was a connection between these variables. The present authors could see no relationship between these same two variables. It may be that Clough's cycle lengths for the early years are erroneous.

length of the periodic changes shown by the spots, the solar activity varying inversely with the length of the 11-year period." Clough is simply saying that weather variations and solar cycle lengths are related.

   Clough's articles did not induce others to follow his line of research. The only other article we have located that discussed solar-cycle lengths and climate is one by Dr. Johannes Muller in 1926; it discusses the parallelism between the return periods of severe winters and longer and shorter sunspot duration peri-

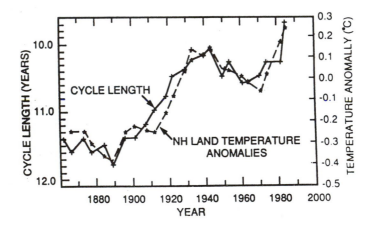

FIGURE 10.13 Variations in solar-cycle length and Northern Hemisphere temperature anomalies. The two plotted variables parallel each other quite remarkably (from Friis-Christensen and Lassen, 1991, with permission). Also note that the solar-cycle length variations are similar in form to Indiana's temperature extreme frequencies as plotted in Figure 4.5.

ods. Muller seems totally unaware of Clough's articles, yet reaches similar conclusions. Another scientist who noticed the apparent relationship between changes in solar-cycle length and climate was J. R. Bray who in 1965 noted more ice off Iceland and advancing Northern Hemisphere glaciers during long solar cycles. The articles of Clough, Muller, and Bray were mostly forgotten soon after publication. In 1991 Eigil Friis-Christensen and K. Lassen of the Danish Meteorological Institute published an article in *Science* entitled "Length of the solar cycle: An indicator of solar activity closely associated with climate." This article attracted considerable attention, probably because the authors presented their results in a better way (Figure 10.13). This paper reaches conclusions similar to those the earlier articles. After showing that the envelope of solar activity makes explaining the variations in Northern Hemisphere temperatures difficult (Figure 10.10), Friis-Christensen and Lassen searched for another solar index that might provide a better explanation. They chose a solar-cycle length index, smoothing the values using a 1-2-2-2-1 weighting of the individual cycle lengths. Figure 10.13 shows that cycle lengths and temperature anomalies closely track each other. Friis-Christensen and Lassen said this suggests, but does not prove, a causal connection between the two phenomena. They choose solar-cycle lengths simply because this solar index provides a good match with Earth's air temperatures. Friis-Christensen and Lassen also hypothesize that solar-cycle variations provide a proxy measure of solar-irradiance changes with an amplitude of 1% over the last 100 years.

Cycle-length variations among solar-like stars are also consistent with the data presented here. Using 18 solar-like stars, Willie Soon and his colleagues at Harvard find that shorter cycle lengths are associated with brighter chromospheric emissions. Their results suggest that for the Dalton Minimum, the sun achieved nearly maximum dimness.

A major advantage of using solar-cycle lengths is that it is easier to measure the cycle length variations than to measure changes in the amplitude of solar activity. Past solar cycle lengths can be deduced from naked-eye sunspot observations, auroral observations, and the ^{10}Be record. This makes possible very long comparisons between solar and terrestrial phenomena. Before undertaking these comparisons, let us examine some other solar indices.

Sunspot Structure and Decay Rates

In 1955 Nordo pointed out that secular changes in sunspot structure and temperatures appear to parallel each other. His results were published in a technical report and did not attract much attention. In 1979 D. V. Hoyt confirmed these results and extended the study to yet another possible sun/climate connection. The umbra is a sunspot's dark central portion. The penumbra is the less-dark surrounding region (see Figure 2.2). The ratio of the umbral areas to penumbral areas, or U/P, is a measure of sunspot structure. Hoyt summed all the umbral areas and penumbral areas each year and found their ratio. A few quiet years had to be omitted because there were so few sunspots that a reliable ratio value couldn't be found. Plotting the values of U/P versus the Northern Hemisphere

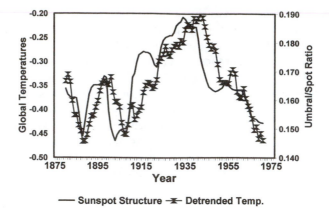

— **Sunspot Structure** —✕— **Detrended Temp.**

FIGURE 10.14 Variations in sunspot structure *(solid line)* and the Northern Hemispheric temperature anomalies *(triangles)* have paralleled each other for more than a century (data from Hoyt, 1979). Both time series are smoothed with an 11-year filter. An upward trend of 0.0043 °C/year in temperature is also removed. This upward trend may be caused by improper analyses (see, for example, Figure 4.2), greenhouse gas warming (see chapter 11), or some other effect.

temperature anomalies tabulated by M.I. Budyko in 1969 and by J. K. Angell and J. Korshover in 1977 produced considerable similarity. Figure 10.14 replots Hoyt's results using more modern temperatures, while smoothing both the U/W values and temperatures with an 11-year filter. A long-term upward trend in temperature is also removed. The 1979 results were statistically significant and would be expected to occur by chance only once for every 10,000 pairs of time series compared.

Like Clough and Friis-Christensen, Hoyt in 1979 hypothesized that this solar index provided a proxy measure of solar-irradiance variations. He deduced that the peak-to-peak solar-irradiance variations measured about 0.38% over the last 100 years. In this same 1979 article, Hoyt related sunspot structural changes to solar-irradiance changes by deducing that if convective velocities increase, the turbulent pressure of the photosphere surrounding the sunspot will increase. Increased inward pressure forces the penumbra to become smaller, while the umbra, being isolated from the surrounding photosphere, remains virtually unaffected. Thus, the U/P ratio will increase because P decreases. Higher convective velocities would be associated with both a brighter sun and with higher U/P values. Convective velocities would vary only a few tenths of a percent over many decades and be difficult to measure directly. This simple model provided a possible explanation for his correlation.

Reexamining this problem in 1993, Hoyt and Schatten devised a better explanation. Rather than using a static model to explain the U/P variations, they invoked a dynamic model. In 1988 Moreno-Insertis and Vasquez showed that sunspot decay rates change secularly by as much as 25% over 50 years. Figure 10.15 illustrates both slow and rapid sunspot decay rates. The U/P or U/W variations can be explained by changes in the rates of sunspot destruction. Umbrae disappear at an approximately constant rate, but the penumbrae disappear faster at some times than others. For example, in the 1930s, individual sunspots were destroyed about 25% faster than they were 50 years earlier. Consequently, when sunspots disappear slowly, the average mean penumbral area is larger, so U/W is smaller. Numerous sunspots consisting only of penumbrae are predicted when sunspots decay slowly (see upper panel of Figure 10.15).

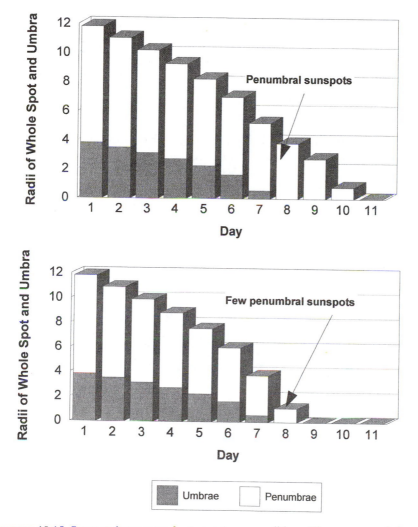

FIGURE 10.15 Sunspot decay rates for two extreme conditions. The upper panel shows slowly decaying sunspots (U/W = 0.163), like those in the late 1800s. Long-lived spots, many penumbral spots, and a low value for U/W (or U/P) are predicted. The lower panel shows rapidly decaying sunspots (U/W = 0.202), like those in the 1930s and today. Under these circumstances, few penumbral spots and a high value of U/W are predicted.

Why are sunspots destroyed? There are at least three possibilities: (1) elements of the sunspot's magnetic flux may be separated from the spot and dispersed, (2) the sunspot may be submerged beneath the photosphere, or (3) magnetic fields of opposite polarity may merge or reconnect, destroying the sunspots. All three of these models for sunspot decay may have one common feature: the rate of sunspot decay is proportional to some convective velocity, either high up in the photosphere (dispersal theory) or deeper in the convection zone (submersion theory). Higher convective velocities mean faster sunspot decay rates and higher values for U/P, as in the static model proposed in 1979. The new theory has one advantage over the older theory—a new prediction. During slow sunspot decay, just before disappearance the remnant sunspot often consists only of a penumbra with no visible umbra. The more slowly the sunspots decay, the more numerous the penumbral sunspots. Hoyt and Schatten predicted that the fraction of penumbral spots should be high in the late 1800s, decrease to a minimum in the 1930s, increase again in the 1960s, and decrease again in recent years. In short, the fraction of penumbral spots should have a shape similar to that of the U/P curve plotted in Figure 10.14. The authors correlated this prediction with Royal Greenwich Observatory measurements from 1874 to 1976 and Rome Observatory measurements after 1958. Figure 10.16 plots their results, and the prediction appears accurate. Since this index

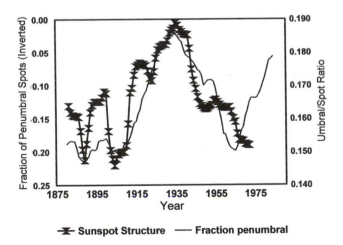

FIGURE 10.16 The fraction of sunspots from 1874 to 1989 shown here have only penumbras. This figure was constructed using 161,714 Royal Greenwich Observatory (RGO) and 24,124 Rome Observatory measurements. The penumbral and total spots were counted for each year. Shown is an 11-year running mean of the number of penumbral spots divided by an 11-year running mean of the total spots. These observations are consistent with sunspot decay and sunspot structure measurements, which are also plotted. Penumbral spots are hypothesized to be more stable when solar convection is weak. The fraction of penumbral spots and sunspot structure have a .77 correlation.

is much easier to measure than U/P ratio, it is probably a better proxy for solar-irradiance changes.

Another plausible, yet unproven, conjecture concerning the connection between sunspot structure, sunspot decay rates, and solar-cycle length is that when individual sunspots decay rapidly, then all sunspots collectively are decaying rapidly. In this case, the entire sunspot cycle tends to disappear more rapidly, and the cycle will be shorter. Variations in solar-cycle lengths are primarily controlled by changes in the cycle's decaying phase length. From this observation, both cycle length and sunspot structure variations may be merely manifestations of changes in the individual sunspot disappearance rate. This explanation may be too simple, however, since it fails to reveal how a new sunspot cycle starts after the old cycle ceases. Nor does it properly account for the rate of new sunspot generation. Despite these two drawbacks, sunspot structure and solar-cycle length variations seem connected by a physical mechanism and not solely by chance.

To summarize, four solar indices with parallel behaviors are solar-cycle lengths, sunspot structure (U/P), sunspot decay rates, and fraction of penumbral spots. Each index has been proposed as a proxy measure of long-term variations in the sun's radiative output. Another solar index, the solar rotation rate, parallels the other indices and may provide the most compelling evidence of all that these indices are proxy measures of solar-irradiance fluctuations.

The Solar Rotation Rate

Several authors have noted changes in solar rotation rates. Because strong coupling between solar rotation and convection is likely, these same authors have suggested that solar rotation can be used as a proxy measure of solar irradiance. The sun rotates fastest at the equator, with one rotation equaling about 25 days as seen from the Earth. Toward the mid-latitudes, the sun rotates in about 28–29 days. These rotation rates are found by watching sunspots traverse the solar disk. In recent years the sun's rotation has been measured both at its surface and within its interior, using omnipresent sound waves. The technique has created a new field of science called helioseismology, which examines various internal pressure waves to deduce the properties of the sun's interior from its surface manifestations. Figure 10.17 illustrates the contours of a constant rotational period expressed in days. The fastest rotation rate is 25 days, located at the equator, while the slowest rates equal 35 days at each pole. The dotted line represents the boundary between the radiative core and the convective envelope. The surface rates map vertically to the core. This rotation configuration differs substantially from theoretical prediction and proved to be a surprise.

Convection is the dominant energy transport mode from the base of the convection zone to the photosphere. This convection can be viewed as blobs or parcels of gas that rise, carrying energy, then dissipate after traveling a certain distance. As they rise, solar rotation carries these parcels from east to west. A parcel rising from a slower rotating region to a faster rotating region, such as along the equator in Figure 10.17, resists the faster rotation. If, for some unknown reason, the solar interior heated up, rising parcels at the equator would travel farther, entering regions of faster rotation. In this case, the equatorial rotation would be expected to slow. Such a slowing equatorial rotation is expected for a brighter sun. It is difficult to imagine a change in solar rotation occurring without some associated change in solar luminosity.

In 1977 Sakurai noted an increasing rate of equatorial solar rotation over solar cycles 18 to 20. In 1974 Eddy and his colleagues found that the solar rotation rate at the beginning of the Maunder Minimum was 4% faster than modern values. Yet in 1981 Abarbanell and Wohl reexamined the drawings of Hevelius used by Eddy and found the solar rotation rate comparable to modern values. Nesme-Ribes (1990) found a slower solar rotation rate in the middle of the Maunder Minimum. Figure 10.18 compares these latter three results with modern values. Finally, in 1990 Hoyt showed that equatorial solar rotation was high in the late 1800s and decreased to a minimum in the second quarter of the 1900s, before once again increasing in recent years (see Figure 10.19).

Theoretically, a strong coupling exists between solar rotation and convection, with convection generally viewed as driving rotation. An observed change

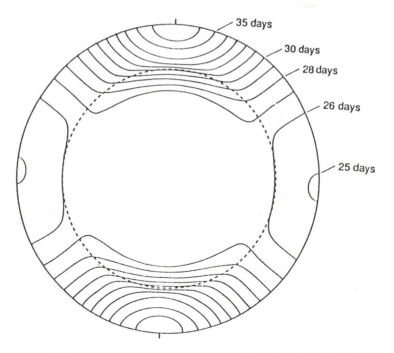

FIGURE 10.17 The sun's interior rotation rates from helioseismology (from Big Bear Solar Observatory, California Institute of Technology, with permission). The fastest rotation rate is 25 days at the equator. The slowest rates equal 35 days at each pole. Each contour is separated by 20 nanohertz or 1.48 days.

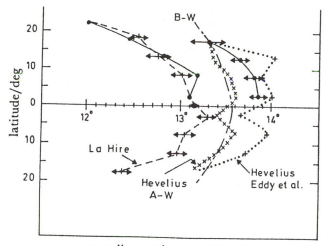

synodic rotation rate/(deg day)

FIGURE 10.18 Three determinations of solar rotation rates in the Maunder Minimum compared with modern values (from Nesme-Ribes, 1990, with permission). Hevelius's observations are at the beginning of the Maunder Minimum, but La Hire's observations are a half century later. Because of conflicting conclusions by Eddy et al. and by Abarbanell and Wohl (A-W in the plot), uncertainty remains.

in solar rotation implies a change in convective energy transport. The following two comments concerning this matter are appropriate: (1) The interaction between convection and rotation is nonlinear and (2) these interactions are strongest in the lower portion of the convection zone. Can solar rotation change without an accompanying change in convection? Any change in solar rotation is a very persuasive indicator that the deeper levels of convection are varying and, hence, there is a corresponding variation in solar luminosity and irradiance.

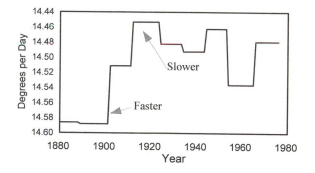

FIGURE 10.19 The solar equatorial rotation rate in degrees per day plotted for the last century. Each solar cycle uses all single sunspots within 60° of the solar meridian and within 5° of the equator.

Calculating the amplitude of solar-irradiance variation associated with a solar rotation variation would be very difficult. Some connection may exist with a change in the number of giant convective cells, as suggested by Nesme-Ribes and Manganey in 1992. In any case, the occurrence of a solar rotation rate variation indicates a solar luminosity change over the past century or a fundamental problem with basics physics. The choice of a solar luminosity change seems easier to accept.

An Empirical Synthesis

Plausible evidence exists for long-term changes in solar irradiance, and we have examined several proposed solar indices for solar-irradiance variations. Proposed proxy measures include sunspot structure, sunspot decay rates, the solar-cycle length, and the equatorial solar rotation rate. Additional proposed proxies include the normalized solar-cycle decay rate and the time rate of solar diameter change. Table 10.1 summarizes the phase relationships among these indices.

In our opinion, variations in the equivalent line widths, the limb darkening of the sun, and line bisectors all indicate possible secular changes in solar convection, the photospheric temperature gradient, and solar irradiance. The variations in these indices may plausibly arise from a common source—secular changes in solar convective-energy transport or convective velocities. Such changes fall outside the domain of usual stellar structure theories, but this is true of all observed solar variations. Without considering any of the arguments put forth in this chapter, it seems implausible that all these solar proxies could vary without any change in solar brightness.

Most solar indices discussed here have been combined to create the solar-irradiance model shown in Figure 10.20. This solar-irradiance model has only two parameters: (1) the amplitude of the solar-cycle variations and (2) the amplitude of the Gleissberg cycle variations. The first parameter is based on the Nimbus-7 observations discussed in chapter 3. The second parameter is chosen so that the Maunder Minimum will be about 0.25% less bright than the modern sun, as suggested by Lean and her colleagues (in 1992), based on stellar variations and the brightness of solar faculae and active network.

Energetically, irradiance variations cannot be ruled out as a source of climatic change. The longer the time scale of the solar variations, the deeper the

TABLE 10.1 Phase relationship of the solar indices.

Index	Year(s) of 20th-Century Maximum
Sunspot structure (U/W)	1934
Fraction of sunspots without umbrae	1933
Rates of sunspot decay	Not available
Solar cycle lengths (1-2-1 filter)	1937.5
Normalized rate of solar cycle decay	1920–1931
Equatorial solar rotation	1924–1934

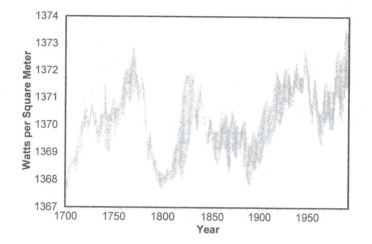

FIGURE 10.20 A plot showing the combined solar-irradiance model using the models based on solar-cycle length, cycle decay rate, mean level of solar activity, solar rotation rate, and fraction of penumbral spots, along with an added solar activity component. The error bars show the relative disagreement among the different empirical models used to derive the irradiance variations. The 1700–1874 data use models based on cycle length, cycle decay rate, and mean level of solar activity. The 1875–1992 data use all five solar indices.

likely perturbation source in the sun. Perturbations in the top few thousand kilometers below the photosphere cause relatively short variations in sunspots and faculae, variations lasting days. Longer variations may arise from deep within the convection zone, perhaps at or just below its base, as suggested from the observed solar rotation changes. Endal and his colleagues in 1985 and Nesme-Ribes and Manganey in 1992 explored possible mechanisms for irradiance variations on the time scale of decades to centuries. Candidate mechanisms include (1) α-perturbations, in which stochastic energy transport variations arise from the finite number of convective cells involved, and (2) β-perturbations, in which changes in pressure, perhaps arising from magnetic field strength variations, alter the rate of energy transport. Plausible arguments can be made on behalf of both mechanisms. Assuming that a perturbation with a radiative flux of 0.2% lasts for one century (see Figure 10.20), this equals one part in 40,000 of the total thermal energy stored in the convection zone. In contrast, during the last century the thermal energy storage in the Earth's atmosphere has varied by about one part in 500.

The correlation between the solar indices and modeled solar irradiance with the Earth's temperature displays a significant confidence level exceeding 99%. Figure 10.20 overlays a plot of the model solar irradiance on a Northern Hemispheric temperature reconstruction from 1700 to the present (see Figure 10.21). Apparently, the sun was dimmest near the Dalton Minimum around 1800, when many regions were very cold. Both 1812 and 1816 were particu-

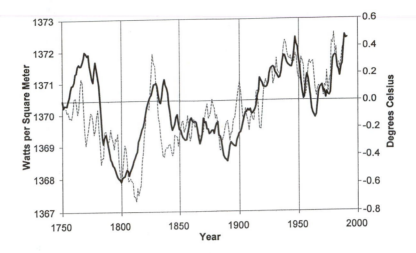

FIGURE 10.21 The annual mean Northern Hemisphere (NH) temperature variations for 1700–1879 from Groveman and Landsberg (1979) and for 1880 to the present from Hansen and Lebedeff (1988), after being smoothed by an 11-year running mean *(lighter line)*. The model irradiances are overlain to show their similarity. Only a slight divergence of the two curves exists for the last few decades.

larly cold, and, in New England, 1816 was called the "year without a summer." Following three out of five cold years after 1816, suffering farmers and others migrated en masse from Maine to Ohio. Perhaps the sun contributed to this prolonged cold period.

Despite the apparent good agreement between the shapes of the two curves in Figure 10.21, a serious problem remains for sun/climate theories. The amplitude of the solar irradiance variations is only about 0.14% from 1880 to 1940. If the Earth's climate responded with a 0.5 °K warming, then the climate would be much more sensitive to solar forcing than most models imply. The direct effect of a 0.14% increase in solar irradiance could account for only about a 0.23 °K increase if the sensitivity to solar influences is 1.67 °K per 1% increase in solar output (based on satellite observations for the last decade). The amplitude of the solar variations remains highly uncertain, with most estimates for the past century ranging from 0.14% to 0.38%. On decade-long time scales, the climate sensitivity is expected to be larger due to several potential positive feedbacks, including a decrease in the ice and snow cover, an increase in plant absorptivity as a warmer world becomes greener, an increase in absorption by water surfaces as wind velocities decrease (based on changes in the length of the day), and changes in plant orientation and albedo as wind velocities vary. These last three potential feedback loops are not included in present-day climate models. Possible problems exist with current climate models if they try to explain the observed correlation. In fact, the variations in solar irradiance would need to be nearly 10 times larger than those shown in Figure 10.21. Perhaps present climate models underestimate the sensitivity of climate to solar variations by failing to properly model the solar spectral variations, the vertical spatial energy input distribution, or the geographical climatic responses.

The postulated solar-irradiance variations may explain a fraction of the Earth's interannual temperature variations, but their effects are most evident in long-term temperature trends on hemispheric and global spatial scales. We now examine some major variations in climate and solar behavior over the last millennium to see if there is any confirmation of a sun/climate connection.

Was the Maunder Minimum Cold?

Evidence shows that during the Maunder Minimum from 1645 to 1715, the climate was anomalous. In the American Southwest, A. E. Douglass's studies of 1919 of tree rings suggested that the late 1600s was a period of drought. Douglass also wrote that during the Maunder Minimum the 11-year tree ring cycle disappeared. Earlier, we described the frequent freezings of the Thames River in the 1660s and 1670s. Gates and Mintz (1975) say that the Maunder Minimum brought unseasonable cold to Europe, with some of the most severe winters of the Little Ice Age, a cool period lasting from about 1430 to 1850. This period is often defined more loosely, with some authors referring only to part of the interval.

In contrast to these positive accounts, Clough said that 1675 was actually a minor warm peak. In 1977 Landsberg and Kaylor pointed out that Tokyo was not especially cold during the Maunder Minimum. By 1980 Landsberg identified the years 1605 to 1615, 1674 to 1682, and 1695 to 1698 as particularly cold; his reconstruction also suggests that the half century before the Maunder Minimum was actually colder. In 1995 Pfister found that the periods 1570–1600, the 1690s, and the 1810s stand out as being cold in Switzerland, as well as in Europe. Scandinavian tree rings, however, show the Maunder Minimum was a period of relative warmth.

All these anecdotal and regional accounts give a confusing and contradictory picture of the Maunder Minimum climate. As pointed out in chapter 4, considerable local variability exists, so any external forcing is difficult to discern. Figure 10.9 shows the expected temperature response to a prolonged solar-irradiance decrease. Regional variations remain sufficiently strong that some areas warm even as the Earth as a whole cools. For example, this model shows that the region just east of Japan may have been warm, which might explain why Landsberg and Kaylor did not find Tokyo particularly cool.

Two items are needed to resolve this problem: (1) a reliable temperature reconstruction of at least hemispheric scale and (2) a plausible solar-irradiance variation model. Fortunately, both are now available. For many years, Ray Bradley at the University of New Hampshire has been assembling and assimilating a variety of climatic proxies. He has reconstructed temperature variations back to 1400, but the results from 1600 on are the most reliable. For the solar-irradiance variations we use solar-cycle lengths suggested by Clough and by Friis-Christensen and Lassen. The work on reconstructing solar activity described in chapter 2 allows us to deduce solar-cycle lengths back to 1610.

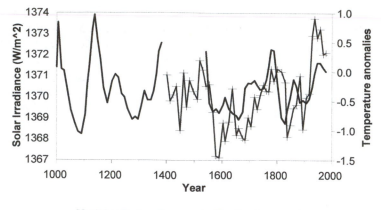

—— Model solar irradiance —+— Bradley temperature

FIGURE 10.22 Solar cycle lengths as given by Youji et al. (1979) and Northern Hemisphere temperature anomalies from Bradley (private communication). Before 1600 both parameters are poorly measured. After 1600 both parameters track each other rather closely, except for the high-temperature excursions in the twentieth century. Could this be the carbon dioxide greenhouse warming effect?

Clough used auroral records to deduce cycle lengths back to A.D. 300. Figure 10.22 plots cycle lengths from A.D. 1000, deduced from Youji and colleagues in 1979, who used Oriental sunspot observations along with Bradley's temperature reconstruction from A.D. 1600. We think most people would agree that these two curves are similar. To parallel one another for four centuries would be quite a coincidence. Temperatures in the 1900s appear higher than expected if solar-irradiance variations were the sole cause of climatic change. Perhaps the carbon dioxide greenhouse effect is manifesting itself. We find that solar-cycle lengths were actually longer before than during the Maunder Minimum. Wolf attributed a 1649 peak to the fact that it occurred 11 years after a well-defined 1638 peak, but he based his claim on only 1 day of observation. Using Hevelius's observations, which Wolf overlooked, we find a solar peak in 1653, suggesting a 14- to 15-year-long cycle at the beginning of the Maunder Minimum. Solar peaks appear evident in 1612, 1626, 1638, and 1653 and produce cycle lengths of 14, 12, and 15 years. These cycle lengths are comparable to the Dalton Minimum cycle peaks. In each case, long cycle lengths are associated with cold temperatures.

Although these discussions have concentrated on temperatures, many other climate variables appear to have solar connections. The sun has an 80–90 year Gleissberg cycle that manifests itself in total activity, cycle lengths, sunspot structure (85 years from maximum entropy analysis), and probably other parameters for which we do not have sufficiently long time series. Cycles of the same length have been reported for many climate and proxy climate parameters. Table 10.2 summarizes some of these cycles. In climate studies, additional reports exist of commonly discussed cycles of this length.

TABLE 10.2 Some meteorological and proxy time series showing cyclic variations close to the 80 to 90-year Gleissberg cycle.

Parameter	Cycle length	Author(s)
Temperature related		
Central England temperature	76 years	Burroughs, 1992
Interdiurnal temperature variability in Budapest	80 years	Berkes, 1955
Prague climate	89 years	Thraen, 1949
Winter severity (Europe)	83 years	Clough, 1933
Winter severity (Europe)	80 years	Maksimov, 1952
Icelandic ice concentration	80 years	Maksimov, 1952
Oxygen isotopes (Greenland)	78 years	Dansgaard et al., 1973
Deuterium isotopes (tree rings)	95 years	Libby, 1983
Precipitation related		
Beijing rainfall	84 years	Burroughs, 1992
Low stage of Nile	83 years	Clough, 1933
Nile floods	77 years	Hameed, 1984
Caspian Sea level	80 years	Maksimov, 1952
Midwestern U.S. droughts	90 years	Moseley, 1944
U.S. drought index from tree rings	90 years	Burroughs, 1992
Tree rings		
Lapland tree rings	90 years	Lamb, 1972
Sequoia growth per decade	83 years	Clough, 1933
California tree rings	80 years	Maksimov, 1952
Tree rings	90 years	Moseley, 1940

Was the Cold Climate of the 1300s Solar Induced?

Cycle lengths can be measured for periods far back in time, and so potentially provide a long history of solar-irradiance variations. It is therefore tempting to look at earlier climate anomalies for a sun/climate connection. Figure 10.22 shows solar-cycle lengths back to A.D. 1000 as well as Bradley's temperature reconstruction to about 1400. Before 1600 the oxygen isotope record may provide a better and more objective measure of global climate. Figure 10.23 shows these two curves in parallel.

Figure 10.23 reveals an inordinately cold interval around 1300. Extremely long solar-cycle lengths imply a dim sun. The ^{14}C record, shown in Figure 10.6, also reached a minimum at about 1350. Since the level of solar activity tends to lag behind the cycle-length variations, this suggests that Youji et al.'s cycle lengths may be correct. These results imply that the 1300s would indeed be cold if a real sun/climate connection exists. In his book *Climatic Changes:*

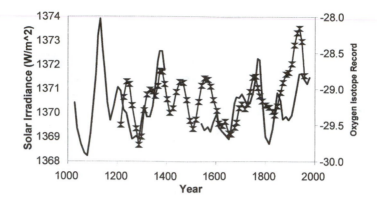

— Naked eye sunspots —✕— O-18/O-16 ratio

FIGURE 10.23 The oxygen-isotope record for solar-cycle lengths from the Dye ice core as given by Youji et al. (1979). These isotopic ratios are proportional to past temperatures. Fair agreement exists between the two tabulations. Even the extremely cold interval around 1300 shows up in both curves.

Their Nature and Causes, published in 1922, Ellsworth Huntington and S. S. Visher point out the anomalous climate of the fourteenth century, whose climatic changes they attributed to solar changes. The 1300s had their share of climatic anomalies, such as:

- During winters, the Rhine, Danube, Thames, and Po rivers froze for months at a time.
- The Baltic froze over in 1296, 1306, 1323, and 1408.
- In 1300 the frequency of severe winters in Europe reaches its second greatest maximum of the last millennium, only to be exceeded around 1600–1650 (see Figure 10.12).
- European floods are recorded during 55 summers in the 1300s.
- Ocean storm surges of unparalleled violence occur in 1300, inundating half the island of Heligoland. In 1304 the island of Ruden was created by being torn from the mainland. On 19 other occasions violent storms took place.
- In 1306–1307 the Caspian Sea was 37 feet higher than at present.
- The early 1300s created prolonged droughts in India, so by the time of the great famine of 1344, not even the Mogul emperor could get food.
- Greenland's weather turned progressively colder, eventually ending the Viking settlements.
- During the 1300s the Nile flood stages were at their lowest levels in the past millennium.
- California Sequoias are growing slowly, implying drought or cold.
- An extended drought from 1276 to 1299 in the western United States

causes abandonment of cliff dwellings and pueblos. Tree rings show severe droughts in 1280, 1283, 1286, 1288, 1295, and 1299.

• In the years just after 1300, England's wheat prices reach their highest value than at any time between 1200 and 1600.

These tabulations suggest a very unusual worldwide climate, with cold winters and wet summers in the northern regions and prolonged and severe droughts in the southern and tropical regions. Low ^{14}C levels and long solar-cycle lengths suggest a cooler sun and, perhaps, a sun/climate connection. This anecdotal evidence does not prove the case, but does indicate that the problem needs more study.

The Solar-Stellar Connection on Long Time Scales

The sun has been observed closely for only three centuries, and its radiant output has been reliably monitored from satellites for 15 years. Although cosmogenic isotopes provide some indication of solar behavior over the last few thousand years, they do not conclusively demonstrate that the sun is a variable star on the time scale of centuries to millennia. Solar-like stars can provide some clues about how the solar-radiant output may change on these time scales. Most stars are variable during short time scales (see Figures 3.19 and 3.20).

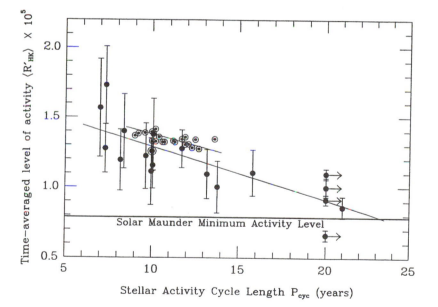

FIGURE 10.24 The level of stellar activity for solar-like stars plotted as a function of stellar-cycle lengths (from Baliunas and Soon, 1995, with permission). Short cycles are associated with higher activity and presumably greater radiant output. The stellar analogues imply that solar-cycle lengths can be used a proxy measure of solar radiant output.

Figure 3.17 hints of longer variability for the star HD 10476 because successive minima had different activity levels. Solar-cycle lengths and climate have also varied in parallel over the last few centuries.

The cycle lengths of many solar-like stars have been measured during the last 30 years. If solar-cycle lengths are proportional to the solar-radiant output, then stellar-cycle lengths should be proportional to stellar brightness. Examining many stellar sources should enable us to estimate solar brightness changes over centuries. A study by Baliunas and Soon in 1995 (see Figure 10.24) found that the level of stellar activity is higher for shorter stellar-cycle lengths. Applying this stellar analog to the sun, Baliunas and Soon found that the sun would be 0.2% to 0.6% dimmer during a grand minimum such as the Maunder Minimum compared with present-day solar activity levels. This estimate is close to Lean's 0.25% value mentioned earlier in this chapter. Solar-like stars all might be variable stars with time scales of decades or longer, which would support the contention that the parallel correlation between solar-cycle length variations and other indices with Earth's temperature is not accidental.

Highlights of the Argument for Solar-Induced Climate Changes

Some highlights of this chapter are:

- Many solar indices vary in parallel with Earth's air temperatures.
- Since 1600, solar cycle lengths and Bradley's climate reconstruction show good agreement.
- Since about 1300, solar cycle lengths and climate based on the oxygen isotope record show moderate agreement.
- Variations in a solar rotation rate may be indicators of changes in solar luminosity due to the coupling between convection and rotation.
- Without long-term measurements to help us understand which solar indices are measures of solar-irradiance variations, many people remain unconvinced of correlative relationships. Such measurements are extremely difficult and may require decades to complete.
- Solar-like stars provide indications that we have observed only a portion of the expected irradiance variation we would have seen had we monitored the sun for many centuries (see Figure 10.24). Solar-like stars vary in brightness on all time scales, so it would be surprising for the sun to be an exception.

In brief, recent studies make a good case that the sun's radiant output varies over decades and longer time scales and that these variations are playing a significant role in climate change.

11. *Alternative Climate Change Theories*

So far, we have primarily considered only the sun's role in natural climatic change. This focus does not imply that the sun is the only cause of climatic changes, nor even the most important one. In the last chapter, we stated that solar and climatic changes have paralleled each other for the last four centuries and, therefore, on time scales of decades to centuries, solar variations might be a dominant driving force for natural variability. Many climatologists adamently disagree with this conclusion and suggest that other factors, both in the past and for the future, are far more important. In the future, mankind's influences may increase and likely will dominate coming climate changes. One hypothesis suggests that the climate system varies randomly, first warming for a few decades and then cooling. Scientists who support this hypothesis believe that external influences need not necessarily cause internal changes. They view the Earth as a thermostat with a very wide interval (or band) of possible temperatures. In other words, the Earth's mean temperature is not constrained to one precise equilibrium temperature. Several natural experiments contradict this belief. For example, following a volcanic eruption the Earth cools for a few months to a few years; then the temperature returns to its preeruption values. This rapid return implies a narrow stable temperature band, with the climate system striving to return quickly to its equilibrium. As the time constant for the climate system is relatively short, it responds quickly to variable forcings. Similar arguments can be made using solar variations. For this reason, we discount the idea that unpredictable chaotic influences completely govern the climatic system.

We will now review three additional forcing functions for climate, specifically volcanic aerosols, anthropogenic aerosols, and greenhouse gas warming. Modern climatologists consider these three forcings the most popular for explaining observed climatic changes. Yet just as the sun/climate connections can be harshly criticized, so can these three ideas, and we will treat these theories with skepticism. We will strive to adopt this same skeptical attitude for solar forcing of climate again in chapter 13.

Volcanic Hypothesis

In the summer of 1783, Iceland's Mt. Lakagigar (Hecla) volcano erupted, spewing ash and dust high into the stratosphere. The westward-traveling winds carried the dust toward Europe, causing widespread comment. Gilbert White, the famous Selbourne, England, naturalist, stated that "the summers of 1781 and 1783 were unusually hot and dry." He further describes the events as follows:

> The summer of the year 1783 was an amazing and portentous one, and full of horrible phenomena; for besides the alarming meteors and tremendous thunderstorms that affrighted and distressed the different counties of this Kingdom, the peculiar haze, or smokey fog, that has prevailed for many weeks in this island, and in every part of Europe, and even beyond its limits, was a most extraordinary appearance, unlike anything known within memory of man. By my journal I find that I had noticed this strange occurrence from June 23 to July 20 inclusive, during which period the wind varied to every quarter without making any alteration in the air. The Sun, at noon, looked as blank as a clouded moon, and shed a rust coloured ferruginous light on the ground and floors of rooms; but was perfectly lurid and blood-coloured at rising and setting. All this time the heat was so intense that butchers' meat could hardly be eaten on the day after it was killed; and the flies swarmed so in the lanes and hedges that they rendered the horses half frantic, and riding irksome. The country people began to look with a superstitious awe, at the red, louring aspect of the Sun; and indeed there was reason for the most enlightened person to be apprehensive; for all the while, Calabria and part of the isle of Sicily were torn and convulsed with earthquakes; and about that juncture a volcano sprang out of the sea on the coast of Norway. On this occasion Milton's noble simile on the Sun, in his first book of *Paradise Lost,* frequently occurred to my mind; and it is indeed particularly applicable, because, towards the end, it alludes to a superstitious kind of dread, with which the minds of men are always impressed by such strange and unusual phenomena.

This one remarkable paragraph, published in 1789, contains nearly all the elements for the hypothesis that explosive volcanic eruptions cause Earth to cool for several months to several years. Explosive volcanic eruptions inject millions of tons of dust and sulfur into the stratosphere. The dust particles rise above the atmosphere's convective layer and are not washed out like lower level aerosols. These aerosol particles reflect light back to space, cooling the Earth. Following these eruptions, the sun as seen from Earth appears dimmer and is surrounded by a large aureole, or bright ring.

Although Gilbert White's haze is not specifically associated with a volcano, the haze is said to be long-lasting and high in the atmosphere because it is unaffected by winds. The red sunrises and sunsets and the dull orb of the sun are clear indications that White was observing a volcanic haze. This account also associates the haze with the great summer heat. During the winter of 1783–1784 a great cold spell in December 1783 produced the coldest frost in 44 years. Snow fell for 2 days in Selbourne. From a modern perspective, this cold period would probably be associated with the Icelandic volcano eruption, but apparently these thoughts did not occur to White.

In the *Proceedings of the Manchester Literary and Philosophical Society* (1785), Benjamin Franklin, who in 1783 was living in Paris while negotiating an end to the Revolutionary War, also commented on the unusual haze:

> During several of the summer months of the year 1783, when the effects of the Sun's rays to heat the earth in these northern regions should have been greatest, there existed a constant fog over all Europe, and a great part of North America. This fog was of a permanent nature: it was dry, and the rays of the Sun seemed to have little effect toward dissipating it, as they easily do a moist fog arising from the water. They were indeed rendered so faint they would scarce kindle brown paper. Of course, their summer effect in heating the earth was exceedingly diminished.
>
> Hence the surface was early frozen
>
> Hence the first snows remained on it unmelted, and received continual additions.
>
> Hence perhaps the winter of 1783–1784 was more severe than any that happened for many years.
>
> The cause of this universal fog is not yet ascertained. Whether it was adventitious to this earth, and merely a smoke proceeding from the consumption of fire of some of those great burning balls or globes which we happen to meet within our course around the Sun, and which are sometimes seen to kindle and be destroyed in passing our atmosphere, and whose smoke might be attracted and retained by our earth, or whether it was a vast quantity of smoke, long continuing to issue during the summer from Hecla, in Iceland, and that other volcano which arose out of the sea near that island, which smoke might be spread by various winds over the northern part of the world, is yet uncertain.
>
> It seems, however, worthy of inquiry, whether other hard winters recorded in history, were preceded by similar permanent and widely extended summer fogs. Because, if found to be so, men might from such fogs conjecture the probability of a succeeding hard winter, and of the damage to be expected by the breaking up of frozen rivers in the spring; and take such measures as are possible and practicable to secure themselves and effects from the mischiefs that attend the last.

These comments more closely resemble the viewpoint of the modern-day volcanic hypothesis than those of White. Neither author's comments on the subject attracted much attention at the time, perhaps in part because White's account appeared in a popular book rather than as a scientific paper and Franklin's account is contained in an obscure journal. A few other obscure and vague earlier speculations also exist which associate volcanoes and climate.

Since Mt. Hecla's eruption in 1783, several other major volcanic eruptions have ejected aerosols into the stratosphere. The largest eruptions were Tambora

FIGURE 11.1 A photograph of the June 11, 1991, eruption of Mt. Pinatubo in the Philippines (from Michael Dolan, with permission). Atmospheric transmission was decreased worldwide for 2 years after this eruption.

in 1815, the Babujan Islands in 1831, Krakatoa in 1883, Katmai in 1912, Agung in 1963, El Chichon in 1982, and Pinatubo in 1991 (shown in Figure 11.1). Other lesser magnitude eruptions have also occurred.

Consider the 1815 eruption of Mt. Tambora in the East Indies. Of all modern eruptions, Tambora is the largest, sending about 40 cubic kilometers of ash into the atmosphere with a plume reaching as high as 40 kilometers. This eruption is famous because in New England every succeeding month had a frost somewhere, creating the "year without a summer." Tambora's eruption is almost unanimously identified as the cause, though this conclusion is problematic. For one thing, many locations worldwide were already cool or cooling before the eruption. As 1816 is near the end of the Dalton Minimum, the less-bright sun may have contributed to the many cool years between 1810 and 1820. From this standpoint, Tambora simply added to the cooling already under way.

Although New England had a cool summer in 1816, it was not cool everywhere. Table 11.1 lists some climate conditions just after the Tambora eruption. These conclusions were deduced by scientists at a conference called "The Year without a Summer? Climate of 1816" held in Ottawa in 1988. C. R. Harington's maps summarize the findings of the 39 conference papers. The table delineates regions with temperatures well below and well above normal. For example, the 1816 summer was hot in the Mississippi Valley and warm in eastern Europe and Japan. The warm and cold areas are insufficient to prove or disprove that volcanoes lead to significant cooling. With any forcing function, the natural local variability is so large that even real global perturbations do not show up everywhere. On the whole, however, the evidence seems to support the idea that volcanic eruptions lead to a cooler climate in the following months and years.

TABLE 11.1 Some climate conditions in 1816 possibly caused by the 1815 Tambora eruption, but perhaps amplified by the low solar activity of the Dalton Minimum with a correspondingly dim sun.

1. Cold summer in New England with frosts every month in northern regions and snow in June; higher crop prices in New York

2. Hot summer in the Mississippi Valley

3. Warm in the Yukon, as indicated by tree rings

4. 1810 to 1820 is the coldest decade of the last thousand years for central western South America, as deduced from the Quelccaya Ice Cap record.

5. About 1°C or more below normal in most of Europe

6. About 1°C above normal in St. Petersburg, Russia

7. Temperatures in India are near normal, but very dry in the north.

8. Cold in Tibet as indicated from the Dundee Ice Cap

9. Above-normal precipitation in China

10. Below-average temperatures in Tasmania and New Zealand

From conference on "The Year without a Summer," Ottawa, 1988.

This distribution of temperature anomalies is similar to the anomalies shown in Figure 10.9 for the Maunder Minimum.

The 1831 Babujan Island eruption casts doubt on the idea that volcanic eruptions are of overwhelming importance to climate. We know the Babujan Island eruption is the most important eruption between 1815 and the eruption of Krakatoa in 1883 because in 1883 many old-timers recalled that the unusual sky conditions in 1831 were similar to those seen in 1883 and 1884. In 1952 Helen Sawyer Hogg summarized those conditions:

> The extraordinary dry fog of 1831 was observed in the four quarters of the world. It was remarked on the coast of Africa on August 3, at Odessa on August 9, in the south of France and at Paris on August 10, in the United States on August 15, etc. The light of the Sun was so much diminished that it was possible to observe its disc all day with the unprotected eye. On the coast of Africa, the Sun became visible only after passing an altitude of 15° or 20°. M. Rozet, in Algeria, and others in Annapolis, U.S., and in the south of France saw the solar disc of an azure, greenish, or emerald colour. The sky was never dark at night, and at midnight, even in August, small print could be read in Siberia, at Berlin, Genoa, etc. On August 3, at Berlin, the Sun must have been 19° below the horizon when small print was legible at midnight.

Despite this remarkable account, when studying the effect of volcanoes on climate, modern climatologists generally omit or discount the Babujan Island eruption. There are good reasons to do so because the following years became steadily warmer worldwide, exactly opposite what one would expect if volcanoes are a major contributing factor to climatic change.

Although the Babujan Island eruption does not fit the volcanic hypothesis at all, most other eruptions do. In a critical study of nine large eruptions during

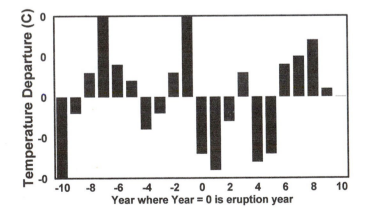

FIGURE 11.2 A superposed epoch analysis (or Chree analysis, Appendix 3) of Northern Hemisphere land temperatures before and after volcuanic eruptions (data from Mass and Porter, 1989, with permission). At the zero year, the explosive volcanic eruption occurs. There is a transitory cooling for 1 or 2 years after the eruption, but this cooling is not much different from other transitory events not associated with volcanic eruptions.

the last century, Mass and Portman in 1989 found some cooling. In Figure 11.2, Mass and Portman superimpose the climatic records just before and after eruptions, with the eruption dates defined as year zero, and plot the average results. Air temperatures cool immediately after the eruption, but 3 years later the climate has returned to normal. Sea-surface temperatures reveal a less distinct signal, and no signal appears in precipitation or surface pressure. Mass and Portman conclude that "only the largest eruptions . . . are suggested in the climatic record, and that modest cooling (about 0.1 to 0.2 °C) is observed for 1 to 2 years after these large events." This statement suggests that volcanic eruptions cause perturbations to the climate, but may not necessarily cause long-term secular trends in hemispheric and global surface temperatures. Some climatologists would argue vigorously against this last idea, but we will not pursue the volcanic hypothesis further.

Increased Anthropogenic Aerosols and the Associated Cooling

The Industrial Revolution began in the late 1700s and by the mid-1800s was dramatically changing English life. Farms were disappearing and farmers were moving to the cities. Coal replaced wood for heating houses. Families burning coal for heat, and cooking in city residential areas created thousands of smoke and ash columns. In London, intense fogs obscured everything, while in the countryside a haze dimmed the sun. Downwind from England, routine North Sea shipboard observations reported lower visibility for this period than for any other time in history.

For many, this time proved traumatic. Poet John Ruskin provides a dramatic description of the effects of increased manmade pollution on the psyche:

> It is the first of July, and I sit down to write by the dismallest light ever yet I wrote by; namely, the light of this midsummer morning, in mid-England (Matlock, Derbyshire), in the year 1871. For the sky is covered with grey cloud;— not a rain cloud, but a dry black veil, which no ray of sunshine can pierce; partly diffused in mist, feeble mist, enough to make distant objects unintelligible, yet without any substance, or wreathing, or colour of its own. And everywhere the leaves of the trees are shaking fitfully, as they do before a thunderstorm; only not violently, but enough to show the passing to and fro of a strange, bitter, blighting wind. Dismal enough, had it been the first morning of its kind that summer had sent. But during all this spring, in London, and at Oxford, through meagre March, through changelessly sullen April, through despondent May, and darkened June, morning after morning, has come grey-shrouded thus. And it is a new thing to me, and a very dreadful one. I am fifty years old, and more; and since I was five, have gleamed the best hours of my life in the Sun of spring and summer mornings; and I never saw such as these, till now. And the scientific men are busy as ants, examining the Sun, and the moon, and the seven stars, and can tell me all about them, I believe, by this time; and how they move, and what they are made of. And I do not care, for my part, two copper spangles, how they move, nor what they are made of. I can't move them any other way than they go, nor make them anything else,

better than they are made. But I would care much and give much, if I could
be told where this bitter wind comes from, and what it is made of. For, perhaps, with forethought, and fine laboratory science, one might make it of
something else. It looks partly as it were made of poisonous smoke; very
probably it may: there are at least two hundred furnace chimneys in a square
of two miles on every side of me. But mere smoke would not blow to and fro
in that wild way. It looks to me as if it were made of dead men's souls—such
of them as are not gone yet where they have to go, and may be flitting hither
and thither, doubting, themselves, of the fittest place for them.

From the mid-1930s to the mid-1970s, the Northern Hemisphere cooled,
and by the mid-1970s this cooling was popularly attributed to man-made aerosols produced by factories and other sources. Theory stated that these aerosols,
sometimes referred to as the "human volcano," scattered incoming sunlight
back to space, causing global cooling. If valid, then atmospheric transmission
would be decreasing in many regions. By the 1950s, we know the London fogs
that Ruskin commented on were disappearing due to antipollution laws. Over
the North Sea there were fewer reports of haze. Presumably, the haze was
increasing at other locations. In Ft. Collins, Colorado, a very large decrease in
atmospheric transmission was evident using an Eppley bulb type pyranometer.
Was the atmosphere becoming hazy and dimming the sun? Closer scrutiny revealed the pyranometer's paint was turning from black to green, reducing the
instrument's sensitivity. Here the instrument was the problem, not the atmosphere. Other locations revealed similar problems.

To circumvent these problems, Howard Ellis at the Mauna Loa Observatory in Hawaii invented the pyrheliometric ratioing technique. Pyrheliometers
track the sun and measure its light output during the course of the day. Ratioing
the output at air mass 2 to 3, for example, produces a number that is independent of the pyrheliometer's calibration. Theoretically, these ratios are sensitive
to changes in atmospheric transmission caused by changes in aerosol loading.
Applying this technique to the Mauna Loa observations revealed that, except
for major volcanic eruptions such as Mt. Agung, over the years the background
transmission remained quite constant.

Perhaps this remote island mountain observatory is too distant to reveal
anything about mankind's aerosol emissions. Fortunately, Mauna Loa is not the
only site where Ellis's technique can be applied. Davos, Switzerland, has records for 1909 to 1979. Davos's atmospheric transmission is remarkably constant (see Figure 11.3), despite being centered in industrial Europe where climatologists predict the maximum signal from man-made aerosols (see Figure
11.4). Climatologists assume that direct measurements of atmospheric transmission, as shown in Figure 11.3, do not exist. In fact, a vast body of measurements is available from 1900 onward for which similar figures could be drawn.
For example, McWilliams showed that transmission changes from 1950 to 1972
in Ireland remained unchanged. For 1951 to 1970, Dogniaux and Sneyers in
1972 uncovered no transmission changes in Belgium.

Despite these negative findings, many modern climatologists try to determine the effect of man-made aerosols by summing up factory emissions, deduc-

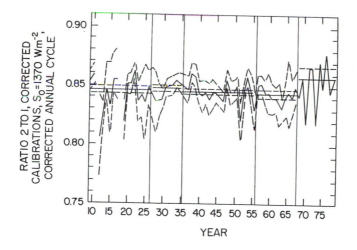

FIGURE 11.3 Atmospheric transmission at Davos, Switzerland, from 1909 to 1979 using the atmospheric ratioing technique. Except for a few large volcanic eruptions, atmospheric transmission at this central European site has remained remarkably stable, with no significant trend. The discontinuity after 1968 arises from a change in sampling during the day and not from any atmospheric alteration. (From Hoyt and Frohlich, 1983.)

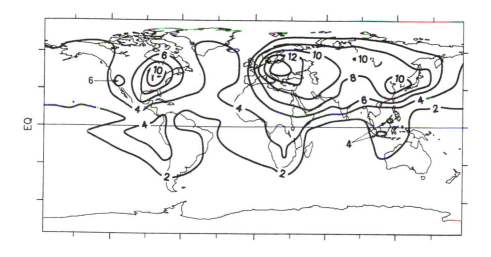

FIGURE 11.4 The spatial distribution of anthropogenic sulphate aerosols, based on model calculations. (adapted from Charlson et al., 1991, with permission). The most intense source regions are Europe and the United States. Nearly all the sulphate aerosols appear in the Northern Hemisphere. Aerosol amounts and their effects are calculated from factory emissions and the use of a chemical and radiative transfer model. They are not based on direct observations, which do not necessarily support these model results. Davos, Switzerland (Figure 11.3), is located very close to where Charlson et al. predicted the maximum effects of atmospheric transmission to occur.

ing the aerosols formed from a chemical model, and then using a radiative transfer model to calculate their effects. Figure 11.4 shows the final net effect, expressed in watts per square meter. This complicated procedure leaves room for many errors, and the calculated effects on the radiation budget of this extra radiation reflected by the Earth due to man-made anthropogenic aerosols range from -0.3 to -1 W/m^2, which is an uncertainty factor of 3. The few direct measurements available suggest the lower limit is nearer reality.

What can we conclude from all these contradictory results? It appears that man-made aerosols have not increased sufficiently to influence climate significantly. Others suggest that these aerosols must exist and ultimately lead to a long-term cooling trend. This same cooling trend is not seen in the climate record because it is balanced by a warming trend caused by increased greenhouse gases such as carbon dioxide. We now examine greenhouse gas warming as a source of possible climatic change.

Greenhouse Gas Warming

In 1896 first Arrhenius and then Chamberlain independently advanced the idea that changes in the amount of carbon dioxide (CO_2) could lead to changes in atmospheric temperature, an idea now called the "greenhouse effect." The process is somewhat akin to a greenhouse in which the transparent glass admits solar radiation but remains opaque to thermal radiation. Incoming solar radiation is absorbed and escaping thermal radiation is hindered, so the greenhouse becomes warmer. Analogously, some solar radiation impinging on the earth is absorbed, heating the surface and atmosphere. This heated material emits thermal radiation, mostly between 7 and 30 microns (see Figure 4.6). The thermal radiation escapes most easily at wavelengths having little atmospheric absorption, such as the region between 11 and 12 microns. At other wavelengths, CO_2 (and, to a greater extent, water vapor) absorbs much of the upward streaming radiation, heating the atmosphere. This absorption warms the entire Earth system. If the concentration of CO_2 increases, the absorption by CO_2 becomes greater and the Earth becomes warmer.

During the International Geophysical Year in 1958, Charles Keeling began measuring atmospheric carbon dioxide concentrations at Mauna Loa. The concentrations are generally expressed as parts per million (ppm) by volume of atmospheric CO_2. Figure 11.5 illustrates the CO_2 concentration as a function of time, showing that the CO_2 concentration has increased from about 315 ppm in 1958 to about 355 ppm today. Additional research has established that about 290 ppm of CO_2 existed in 1880, and perhaps about 275 ppm before the Industrial Revolution started in 1800. The increase is caused by the burning of fossil fuels such as coal and oil.

The CO_2 increase will create a warmer Earth, but by how much? In climatology, the greenhouse warming debate has probably generated more scientific articles than any other topic. Many studies attempt to show how much warming would occur if the CO_2 concentration doubles. The numbers usually range from about 1 to 6 °C, with most values falling in the range of 2 to 3 °C. If, for the

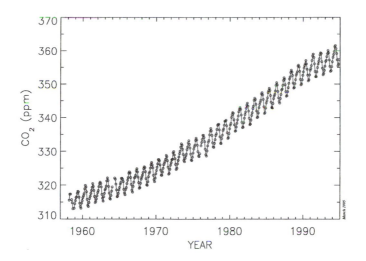

FIGURE 11.5 Monthly mean concentrations of carbon dioxide at Mauna Loa since 1958 (from Pieter P. Tans, personal communication, with permission). Both locations show a steady increase due to the burning of fossil fuel. Mauna Loa has a strong annual cycle caused by the seasonal growth and death of Northern Hemisphere plants.

moment, we assume that a CO_2 doubling causes a 2.5 °C warming, then the increase from 290 ppm in 1880 to 355 ppm in 1992 would cause a 0.73 °C warming. This number can easily be calculated using the following formula:

$$\text{Warming} = \frac{2.5\ln(355/290)}{\ln(2)} \quad (7)$$

in which 2.5 is the warming in degrees Celsius for a CO_2 doubling, 290 and 355 are the beginning and ending concentrations in ppm, and $\ln(2)$ is a normalization for the concentration doubling.

The predicted 0.73 °C warming is very close to the observed 0.6 to 0.7 °C warming of the Northern Hemisphere in the last century. This might lead to the conclusion that the observed climatic variations could be explained by a greenhouse warming, with perhaps some random variations centered around the long-term increase. Countering this argument is the observation that most of this warming occurred before 1930, before greenhouse gas concentrations increased significantly. In addition, long-term changes in solar irradiance can explain most observed changes in climate during the last several centuries. Yet CO_2 must have some role in climatic change. A voluminous literature has developed trying to quantify the greenhouse effect.

Because the expected greenhouse warming has not proven as strong as most theoretical calculations predict, some scientists suggest that the greenhouse effect is masked by increased anthropogenic aerosols. Yet some limited observational evidence suggests that these anthropogenic aerosols are actually becoming more plentiful. Considerable debate continues over what factors cause climatic change. Certain scientists seek confirmation of a favored theory while discrediting other theories. In fact, the sun, greenhouse gases, volcanic aerosols, and perhaps anthropogenic aerosols may all be playing roles. The next section examines many pieces of this puzzle in an effort to determine the importance of each component in causing climatic change.

Putting the Pieces Together

Each forcing function described in this chapter plays a role in climatic change. From the record, we believe the sun plays a major role in natural secular climatic changes on time scales of decades to centuries. The sun also contributes to a weak 11-year cyclic variation. The moon causes an 18.6-year oscillation. Explosive volcanic eruptions create weak cooling 1 to 2 years after their occurrence. Increasing CO_2 and other greenhouse gases are producing a long-term secular warming that, by the middle of the next century, may be the dominant cause of climate change between now and then. Anthropogenic aerosols may

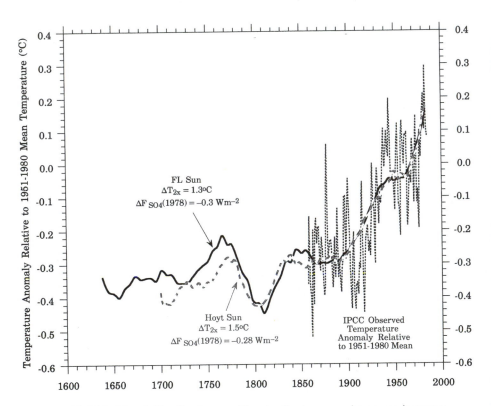

FIGURE 11.6 Models of climatic change with solar forcing, greenhouse gas increases, and anthropogenic aerosol increases (from M. Schlesinger, personal communication). Two solar forcing models are used, one by Hoyt and Schatten (1993) and one by Friis-Christensen and Lassen (1991). Both models indicate a greenhouse gas warming in the range of 1.3 to 1.5 °C.

be causing small, localized regional climate changes. While some meteorologists may condemn these views, many opinions concerning climatic change would likely be denounced from some quarter.

During the last century, several modeling studies have tried to deduce the roles of the sun and greenhouse gases. Figure 11.6 shows one such study by Schlesinger (1994). Schlesinger uses three climate forcing functions: (1) the solar forcing we discussed in the last chapter, (2) a cooling due to anthropogenic aerosols, and (3) a warming due to greenhouse gases. Employing an energy-balance climate model, Schlesinger deduces each component's most likely contributions. He states that a CO_2 doubling would lead to a 1.5 °C warming, leading to the conclusion that, since 1880, CO_2 has created a possible 0.44 °C warming. If half the observed warming is due to erroneous measurements, the CO_2 warming required for a CO_2 doubling would be reduced to 0.75 °C. If the anthropogenic aerosols are not causing any cooling, then the 0.75 °C value will have to be reduced even more. These uncertainties in theory and measurement indicate the need for more study. Comprehensive theoretical and experimental studies may help reduce a multitude of opinions and lead towards a comprehensive theory for climatic change.

12. *Gaia or Athena?*

The Early Faint-Sun Paradox

Stellar evolution theory predicts large, long-term solar luminosity ($L_\odot$) changes over the lifetime of the sun. The most certain prediction is a general monotonic increase (neglecting short-period variations) in $L_\odot$ of about 30% over the past 4.7 billion years, an increase that will continue. This prediction is well founded theoretically (based on the conversion of hydrogen into heavier elements) and supported observationally by the famous Hertzsprung-Russell diagram showing stellar evolution (see Figure 3.18).

If the solar luminosity increases monotonically with time, one might expect to find evidence of increasing surface temperatures in the Earth's paleoclimatic record. Instead, isotopic indicators show Earth's mean surface temperature is now significantly lower than it was 3 billion years ago. In 1975, R. K. Ulrich termed this the "faint young sun" paradox. Simultaneous solar luminosity increase and terrestrial temperature decrease imply additional strong influences on climate evolution. To understand climate evolution (and, by inference, the present climate), we must first determine the nature of these "compensatory mechanisms."

The positively increasing line in Figure 12.1 shows the evolution of solar luminosity (in units of present luminosity, L). Since terrestrial surface temperatures have remained nearly constant during the last 2.3 billion years, this requires a very effective compensatory mechanism.

Several theories attempt to explain why the Earth's surface temperature has remained relatively constant even while the solar luminosity has increased by 30%. Also, various scenarios have been advanced to explain why the Earth remained ice-free even during periods when the sun was much dimmer than it is today. Some of these ideas are:

- Since it had fewer continents and more oceans, the early Earth was much darker. This same darker surface absorbed enough additional incoming solar radiation to remain ice-free.

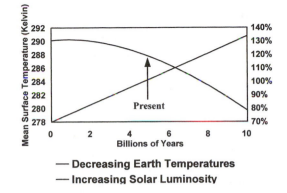

— **Decreasing Earth Temperatures**
— **Increasing Solar Luminosity**

FIGURE 12.1 Predicted solar luminosity as a function of time (data from Schatten and Endal, 1983). The positively increasing line shows solar luminosity has grown by about 30% during the last 5 billion years and will continue to increase in the future. The decreasing line shows a simple model for the Earth's temperature versus time, assuming the atmosphere becomes thinner by 1 millibar (mb) every 10 million years. The present-day sea-surface pressure is 1013.25 mb. In the future, the Earth's atmosphere may boil off into space or be taken up by plants, diatoms, or nonbiotic processes. In contrast, large extinct flying creatures such as pterodactyls and 1-foot-wide dragonflies would be more stable flyers in a postulated earlier thicker atmosphere. There are many other alternative theories to explain the Earth's climate stability.

- In the past, energy transport from the equator to polar regions was easier because the continents had lower elevations. This enhanced heat transport allowed the Earth to remain relatively warm.
- The early atmosphere had more carbon dioxide and methane, creating an enhanced greenhouse effect sufficient to trap the incoming solar radiation and keep the Earth warm. The enormous amount of carbon trapped in limestone suggests that Earth's former atmosphere contained much more carbon dioxide than it does today.
- The early atmosphere was thicker. A loss of only 1 millibar of atmosphere every 10 million years would be sufficient to keep Earth's temperature near its present level. The atmosphere could be lost by boiling off into space, it could be stored by plants or animals, or it could be removed by nonbiotic processes such as rusting. During this scenario, illustrated in Figure 12.1, the Earth is warmer without requiring any other changes. It is analogous to the increased warmth one encounters from going from a mountain to a valley where the atmosphere is thicker. Today's mean sea-level surface pressure is 1013.25 millibars.
- Organisms collectively controlled the Earth's albedo to make it warmer and more optimal for life, a portion of those ideas known as the Gaia Hypothesis. We now examine this concept.

The Gaia Hypothesis

The Gaia Hypothesis, proposed by Lovelock in 1979, has attracted considerable attention. Lovelock is one of those rare, self-supporting scientists not employed by some large establishment, and so has a freedom that has allowed him to develop and pursue ideas that would not otherwise be possible. Lovelock lives in an English countryside cottage and specializes in atmospheric chemistry. He invented the gas chromatograph that accurately measures the concentration of trace gases. When the Jet Propulsion Laboratory was preparing to send a probe to search for life on Mars and measure that planet's atmospheric composition in the mid-1960s, Lovelock, as an expert in this field, was invited to participate and eventually concluded that a planet with life would have an atmosphere different from a planet without life. The Earth's atmosphere has oxygen, methane, and a number of other highly reactive gases. If life were suddenly to cease, these gases and the atmospheric composition could change. Lovelock argued that life controls the composition of the Earth's atmosphere, which is merely an extension of the biosphere. If life existed on Mars or any other planet, it would create an atmosphere that is not in chemical equilibrium. Life could be detected on other planets simply by analyzing that planet's atmospheric composition. An atmosphere not in chemical equilibrium or one with reactive gases would show life existed on that planet.

As originally formulated, the Gaia-Hypothesis stated: "This postulates the physical and chemical condition of the surface of the Earth, of the atmosphere, and of the oceans has been and is actively made fit and comfortable by the presence of life itself. This is in contrast to the conventional wisdom that held that life adapted to the planetary conditions as it existed and they evolved their separate ways."

Lovelock applied this line of thinking to the composition of the Martian atmosphere and concluded that Mars had no life. The same could be said for Venus. Apparently anticipating Lovelock, the astronomer St. John had already reported in 1928 that Venus's lack of oxygen showed no life existed there.

Other people had also considered the atmosphere's composition and its interaction with life. In a particularly charming quotation from "A Year among the Trees," first published in 1881, Wilson Flagg states:

> The ignorant, from want of knowledge, are always theorists; but genius affords its possessor, how small so ever his acquisitions, some glimpses of truth which may be entirely hidden from the mere pedant in science. My philosophic friend, a man of genius born to the plough, entertained a theory in regard to the atmosphere, which, although not strictly philosophical, is so ingenious and suggestive that I have thought an account of it a good introduction to this essay. My friend, when explaining his views, alluded to the well-known fact that plants growing in an aquarium keep the water supplied with atmospheric air—not with simple oxygen, but with oxygen chemically combined with nitrogen by some vital process that takes place in the leaves of plants. As the lungs of animals decompose the air they inspire, and breathe out carbonic acid gas, plants in their turn decompose this deleterious gas, and breathe out pure

atmospheric air. *His theory is that the atmosphere is entirely the product of vegetation,* and that nature has no other means of composing it; that it is not simply a chemical, but a vital product; and that its production, like its preservation, depends entirely on plants, and would be impossible without their agency. But as all plants united are not equal in bulk to the trees, it may be truly averred that any series of operations or accidents, that should deprive the earth entirely of its forests would leave the atmosphere without a source for its regeneration.

Although some elements of the Gaia Hypothesis were already present both within the scientific community and among lay people, Lovelock's interesting new contribution was to contend that the atmosphere is controlled in the most beneficial manner to life collectively rather than individually. Lovelock argued that the faint early-sun paradox did not occur because life collectively controlled the Earth's temperature to create optimal conditions for its own continued existence. Figure 12.2, from Lovelock's book *Gaia, A New Look at Life on Earth,* illustrates his thinking.

Lovelock invented the concept of Daisy World to show how the Gaia Hypothesis might work. Daisy World has two varieties of flowers—a white daisy and a black daisy. When the sun is young and dim, the planet has many black daisies, so more solar radiation is absorbed and the planet is warmed to be conducive to life. As the sun ages and becomes brighter, more and more white daisies reduce the absorbed radiation. The black and white daisies adjust their relative numbers so the amount of absorbed solar radiation remains constant and the planet has a stable temperature over time. Daisy World's biota regulate the planetary temperature much like a thermostat. According to Lovelock, Earth may very well be acting like Daisy World, although the thermal regulation requires more numerous and complicated processes.

Philosophically, the Gaia Hypothesis is very appealing. Over long time scales, comparable to those involved in the evolution of new species, or many millions of years, an active process may control the planet. Clearly, other mechanisms such as the Milankovitch Hypothesis, which states that ice ages are controlled by changes in Earth's orbit, must dominate climatic changes for thousands of years. For example, the tilt of the Earth's axis varies over thousands of years. When the Northern Hemisphere tilts away from the sun during the summer months, snow and ice start building up on Baffin Island and gradually become an ice age. The Milankovitch Hypothesis is relatively simple and thus less controversial than many other climate change theories. On very short time scales, climate can be viewed as chaotic, being dominated by weather systems. Between the chaotic variations and the ice ages, other factors may potentially dominate climatic change. The Athenian Hypothesis describes climatic variations in this intermediate time domain.

The Athenian Hypothesis

The intimate correlation between life and the atmosphere is not a new idea. In a 1869 lecture at University College, London, poet and author John Ruskin

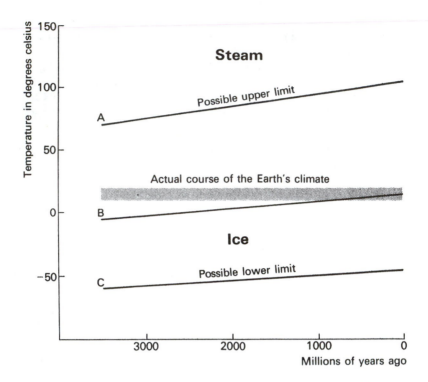

FIGURE 12.2 An illustration by James Lovelock (1979, with permission). We quote his caption: "The course of the Earth's average temperature since the beginning of life 3.5 eons ago is all within the narrow bounds of the horizontal lines between 10° and 20°C. If our planetary temperature depended only on the abiological constraints set by the Sun's output and the heat balance of the Earth's atmosphere and surface, then the conditions of either the upper or the lower extremes, marked by the lines A and C, could have been reached. Had this happened, or even if the middle course were followed, line B, which passively goes with the Sun's heat output, all life would have been eliminated."

FIGURE 12.3 Possible view of the approximate time and space domains where various hypotheses for climatic change have their greatest likelihood for validity. The divisions between the domains of validity are not sharp lines as drawn, but rather they merge together gradually. These divisions are estimates only. In this book we are primarily concerned with the region covered by the Athenian Hypothesis which includes possible solar-irradiance variations.

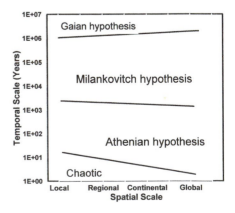

stated: "Athena is the air giving vegetative impulse to the earth. She is the wind and the rain, and yet more pure air itself, getting at the earth fresh turned by the spade or plough, and, above all, feeding the fresh leaves; for though the Greeks knew nothing about carbonic acid, they did know that trees fed on the air." This quotation suggests that the Gaia Hypothesis was already known more than 100 years ago, but the Greek goddess associated with it was Athena, not Gaia. In Ruskin's honor, as a companion to the Gaia Hypothesis, we now introduce the somewhat different Athenian Hypothesis.

The Athenian Hypothesis may be defined as climate having a fixed point that only varies significantly if externally forced. The external forcing might be volcanoes (seasons to years) or variations in solar radiative output (years to millennia). This definition does not include Lovelock's rosy conjectures.

Two other relevant hypotheses are (1) the Chaotic Hypothesis in which climate varies quasi-randomly even with constant external forcing and (2) the Milankovitch Hypothesis in which ice ages are caused by changes in Earth's orbit, such as the eccentricity or the inclination of the rotational axis.

Figure 12.3 illustrates the time and space domains in which the Gaia Hypothesis, the Milankovitch Hypothesis, the Athenian Hypothesis, and the Chaotic Hypothesis may dominate. Logical considerations alone usually do not determine whether one accepts or rejects any of these hypotheses for climatic change. The Gaia Hypothesis has often been criticized. Earth's atmosphere may vary without regard to the existence of life, but surviving life forms find these unregulated changes optimal. Philosophical outlooks, political beliefs, and past experiences all play important roles when accepting or rejecting these hypotheses.

Some Final Words

This chapter explored some ideas associated with climate, its variability and stability. The sun has brightened by about 30% during the last 4.7 billion years. Earth's temperature has remained nearly constant over this same period. Several scenarios can be devised to explain why the temperature remains steady while the Earth receives more and more solar radiation. The Athenian Hypothesis also includes possible solar-irradiance variations.

13. *Final Thoughts*

Some Ironies of Sun/Climate Relationships

In 1872 Meldrum found an apparent relationship between the number of Indian cyclones and the solar cycle. If valid, this strong relationship has significant social and economic impacts. Shipping disasters might be avoided, agricultural yields predicted, and famines prevented. The implications were so important that Meldrum's observation inspired several programs to make detailed solar observations. In 1874 the Royal Greenwich Observatory started detailed observations of sunspot areas and positions, a program that continued until 1976. Ironically, in 1976 Jack Eddy published his paper "The Maunder Minimum" that greatly revived interest in sun/climate studies. Yet just when the best sunspot position and area measurements could be made and analyzed with modern spacecraft data, making them more essential than ever, this measurement program ceased.

For more than 50 years, Charles Greeley Abbot measured the solar irradiance from mountain tops around the world. Despite his best efforts, removing atmospheric effects from his measurements proved virtually impossible. Separating solar changes from atmospheric changes to find the actual solar behavior proved extremely difficult. Only satellite measurements could fulfill Abbot's dream of truly accurate solar-irradiance monitoring. Abbot's dream became reality in 1978 with the launch of the Nimbus-7 satellite. Unfortunately, Abbot died in 1973 at the age of 101.

The Case against Solar Forcing of Climatic Change

Many controversies and ironies exist in the sun/climate field. One additional irony worth mentioning is the claim by Dove and Maury in the last century that the sun's radiant output remains constant and therefore plays no role in climate change. What is surprising about these findings is that although their work influenced other scientists, these men were not solar physicists but ocean-ographers, yet they felt free to define solar behavior. Solar physicists have never been highly confident about the sun's constancy, since the sun shows some variability wherever it is observed.

The first argument points out that the sun's brightness varies by a few tenths of 1% over the solar cycle, and virtually all present-day climate models suggest that the effects on the Earth will be small. At best, Earth temperature changes will amount to only a few hundredths of a degree and will be difficult to detect. Empirical studies tend to support this argument, because 11-year sig-nals are seldom discerned in the climate records.

A second argument against solar forcing derives from the observation that correlations of temperature and other meteorological variables with solar activ-ity often disappear or change sign. These breakdowns suggest that the correla-tions are accidental and have no physical basis.

A third argument against solar forcing of climate states that while small 11-year variations in solar output may be occurring, no evidence exists for any longer term changes in the sun's output. Although several solar behavior mod-els have been advanced, they lack credibility until solar-irradiance measure-ments have been made for many decades. Even the models postulated here prove problematic. For example, the weather was cooler before the Maunder Minimum than during it, suggesting that the envelope of solar activity, which is a proxy for solar-irradiance variations, cannot completely explain climate changes. A large degree of enhanced warmth in the twentieth century occurred during the 1930s, while solar activity did not peak until the 1950s. This can be taken to provide further evidence that there is no sun/climate relationship.

A fourth argument asserts that no physical mechanism can explain any postulated long-term changes in solar brightness. Without a theory, there is no basis to believe such variations exist.

The four arguments against solar forcing of climate are:

- Insufficient changes in solar brightness
- Correlation breakdowns and sign changes
- Terrestrial changes that often occur before changes in the envelope of solar activity and indicate that long-term solar forcing is not in-fluencing climate change
- The lack of solar terrestrial theories to account for any postulated long-term changes

These arguments have led many to believe the sun plays no role in climate change. We now turn to the reasons for maintaining a positive view.

The Case for Solar Forcing of Climatic Change

Two high-confidence-level conclusions concerning the sun's irradiance varia-tions are (1) the solar radiant output varies with an 11-year cycle, with a brighter sun when the sun is more active, and (2) over the last 4.7 billion years the sun's luminosity has increased by 30%, or 0.0064% per million years. While not yet proven, it seems probable that variations occur on intermediate time scales. An 80- to 90-year solar-irradiance variability likely exists because this variability is seen in solar activity. As a result of activity-related variations, compared with today, the sun may have been about 0.25% dimmer during the Maunder Minimum and other grand minima. Some convincing evidence for variations on this time scale comes from measured changes in the solar rotation rate. Variations in solar convection, possibly related to magnetic features, could cause rotation rate and solar-irradiance changes. Changes in sunspot decay rates, revealed as changes in sunspot structure, provide additional evidence for fluctuations in the solar irradiance.

Many meteorological parameters vary with different periods, including the 11-year solar cycle, but most variations are small and difficult to detect. Widespread changes in phase between these cyclic variations and the corresponding solar activity cycles also cause confusion. Some problems existed in sun/climate relationships in some years before 1920. For these years, atmospheric temperatures seem lower when the sun is active than when it is quiet, exactly the opposite of the behavior generally thought to occur. A paradox also arises when considering other meteorological measurements. Precipitation, pressure patterns, the number of tropical cyclones, and so forth, measured before 1920 are all consistent with a brighter, active sun. Why are temperatures inconsistent with all these other parameters and with modern solar behavior? Perhaps the temperatures are not global, but concentrated in regions that today show variations 180° out of phase with what is expected. Currie showed that the western United States follows this pattern today. Perhaps Europe followed this anomalous pattern in the last century. No clear answer explains this, although it may be as simple as natural nonsolar climate variability superposing an out of phase signal. Nevertheless, understanding these phase reversals seems important to offset criticisms. That the 11-year cycle for all weather elements, although weak, is stronger than one would expect based on climate models, also remains unexplained. This abnormal strength is most evident in the nineteenth century.

On time scales of decades to centuries, since 1600 several solar indices track terrestrial temperatures quite well (see Figure 10.19). In particular, solar-cycle lengths seem to be a good proxy for solar-irradiance variations. Temperature change variations for the detrended northern and southern hemispheres are well correlated, suggesting a common source. Solar-irradiance changes may be part of the control mechanism for climatic variations during the last 100 years, if not longer. Both older solar indices and terrestrial climate proxy records are required to confirm whether this correlation has always existed. Accurately reconstructing past solar and climate records is difficult.

The arguments favoring a solar role in climatic change can be summarized as follows:

- Changes in solar brightness on the 11-year time scale are firmly established from satellite measurements. Presumably, there must be some nonzero 11-year climate response to this forcing.

- Discoveries have shown some correlations between solar activity and a large number of meteorological and related variables.
- Terrestrial temperature changes on time scales of decades to centuries display correlations with changes in solar-cycle length, sunspot structure, solar rotation, and other variables.
- Changes in sunspot decay rates (and hence sunspot structure) and in the solar rotation rate may be explained by changes in solar-convective velocities and patterns suggesting changes in the solar constant.
- Variations in solar-cycle length and terrestrial temperatures parallel each other since at least A.D. 1300. Observations of stellar-cycle lengths and stellar activity support associating short cycle lengths with a brighter star.

Should these hypothesized long-term variations in solar irradiance prove true, they would have important implications not only for understanding climatic change but also, possibly, for predicting future climatic changes. This, in turn, could help determine the strength of mankind's influence on the greenhouse effect. Natural variations must be understood before mankind's impact can be separated out. These inquiries will include more theoretical solar and climate models, as well as accurate solar and atmospheric observations of significant parameters to feed into these models.

Solar Forcing of Climatic Change

In this book we have evidenced centuries of study searching for a definitive answer to the question, "Does the sun affect Earth's climate?" Although the question appears simple, we hope that the reader is left with no definitive answer to this interesting question. There are many confusing issues that plague each response. We hope this book has brought more than this confusion to the attention of the reader; our wish is to also convey the zeal for this research!

IV. APPENDICES

Appendix 1. Glossaries

Solar and Astronomical Terms

An *astronomical unit* is the average distance between the Earth and the sun, about 150 million kilometers, or 93 million-miles.

Aurorae, or northern lights, are bright regions caused by the interaction of the solar wind with the Earth's magnetic field high in the atmosphere over northern regions and seen mostly when the sun is active.

The *chromosphere* is the region of the solar atmosphere just above the photosphere. It is the region where many spectral lines are observed; temperatures rise in this region with increasing radius.

The *corona* is the sun's tenuous outer atmosphere which has a temperature of many millions of degrees. The corona has its greatest extent during solar maxima.

Coronal mass ejections (CMEs) are regions often overlying active regions, which eject streams of electrically charged particles that affect the Earth's magnetic field.

Cosmic rays are energetic ionized nuclei and electrons in space and are both galactic and solar in origin. The most energetic are of galactic origin.

Faculae (Latin, bright torches) are bright regions in the photosphere (or visible surface) of the sun associated with magnetic fields; their structure is thought to consist of mini-wells with hot walls.

Flares are brief increases in brightness of small solar regions, probably caused by the breakdown of a magnetic containment field and the subsequent release of mass and light. Flares, which are usually observed in the hydrogen alpha emission line, yield intense bursts of ultraviolet radiation and, on rare occasions, can be observed in white light. Coronal mass ejections are somewhat related to active regions and flares.

Fusion of hydrogen nuclei protons into helium nuclei creates the primary source of solar energy. During this process a small amount of mass is transformed into energy. Nuclear fusion occurs in the very hot (15,000,000 °K) and dense interior of the sun. Although the sun loses half a million tons of matter every second, it will maintain its present energy output for at least another 5 billion years.

Geomagnetic storms, or sudden ionospheric disturbances and radio fadeouts caused by solar active regions and coronal mass ejections, often disrupt telecommunications on the Earth. In polar regions, geomagnetic storms are often accompanied by aurorae or northern lights.

The *Gleissberg cycle* is an 80- to 90-year quasi-periodic variation in sunspot number and other solar indices.

The *Group Sunspot Number* is a measure of solar activity that equals 11.93 times the number of sunspot groups. On average, the Group Sunspot Number is scaled to equal the Wolf Sunspot Number.

The *Hertzsprung-Russell (or HR) Diagram* plots stellar brightness versus stellar color or surface temperature. Most stars cluster together in discrete groups, and the majority of stars fall along a line known as the *main sequence.* Color classifications follow the OBAFGKMN sequence from blue to red known as the *Draper,* or *HD, system.* The sun is a yellow star (class G), about two-tenths of the way on the G scale (hence the number 2), located on the main sequence (hence the Roman numeral V) (see Figure 3.20). Therefore, the sun has a color classification G2V and an effective surface temperature of 5780 °K.

Luminosity is the omnidirectional integrated light output or flux of the sun or other star expressed in units of power such as watts. Strictly speaking, it includes all energy outflows.

The *Maunder Minimum* is the prolonged sunspot minimum from 1645 to 1715. During this interval many consecutive years passed when no sunspots were observed although other solar activity indices may have varied. Other grand minima include the Dalton or Modern Minimum (1795–1823) and the Spoerer Minimum (1450–1540).

The *penumbra* is a sunspot's dark region surrounding the even darker umbra.

The *photosphere* is the sun's visible surface which has a temperature of about 5780 °K.

Plage (French, beaches) are broad diffuse bright chromospheric regions (hence the name) seen in monochromatic light (such as the calcium II K emission line). Plages and faculae overlap each other but have different areal coverages.

The *Schwabe cycle* is the sun's 10- to 11-year cycle, also known as the solar cycle or solar activity cycle.

Solar activity refers to sunspot number, faculae, plages, the corona, and other solar phenomena displaying the 10- to 11-year cyclic pattern associated with magnetic variations.

The *solar constant* is the sun's total irradiance at 1 astronomical unit expressed as the amount of radiant power crossing a unit area. Here, we usually use the terms *solar irradiance* or *solar total irradiance* because it is not strictly a constant.

The *solar convection zone* is a region extending from about 0.70 of the sun's radius to the sun's surface, where convection is the dominant method of energy transport.

Solar granulation is the sun's convection pattern that manifests itself as convection cells called granules. Granules are roughly 1000 km (600 miles) in diameter. The convection pattern has often been likened to the appearance of rice grains floating on the surface of water.

Solar luminosity is the sun's total omnidirectional energy output. Although neutrinos carry away energy when escaping from the sun, the solar luminosity usually refers only to the electromagnetic radiative output.

A *solar sector* is a region of the solar wind with one polarity magnetic field. Typically, the solar wind will be divided into two to four sectors. Boundaries between sectors are called *solar sector boundaries* and become key dates in some sun/weather studies when they cross the Earth.

The solar wind is a low density proton-electron gas streaming from the sun. At the Earth, the solar wind has a mean velocity of 400–500 km/second and a density of 5 particles/cm^3.

Sunspots are dark regions in the sun's photosphere usually consisting of a dark umbra surrounded by a less-dark penumbra. Sunspots vary in number and mean latitude with an 11-year cycle. They have strong magnetic fields and their centers are depressed from the level of the photosphere, an effect known as the Wilson Depression.

Sunspot structure is defined either as the ratio of the umbral to penumbral areas (U/P) or as the umbral to whole spot areas (U/W). Over decades this ratio can vary by as much as 25%, and it is related to the rate of sunspot decay.

The *Wilson Depression* is the difference in depth between a sunspot's visible surface and the surrounding photosphere.

The *Wolf Sunspot Number* is a measure of solar activity equal to 10 times the number of sunspot groups plus the number of individual sunspots multiplied by an observer correction factor (see Group Sunspot Numbers).

The *umbra* is a sunspot's dark central region. Its temperature is about 4000 to 4500 °K, and it only appears dark because the surrounding photosphere is significantly brighter.

Terrestrial Terms

The *adiabat* or adiabatic lapse rate is the vertical temperature-pressure profile of an atmosphere in which rising air parcels change their volume without exchanging heat with the surrounding atmosphere. When the temperature profile is steeper than an adiabat, motions tend to transport energy outward, and vice-versa.

The *albedo* is the fraction of incident solar radiation that is reflected directly back to space without being transformed into heat. Earth's albedo equals about 30%.

Earth's *atmosphere* consists of 77% nitrogen, 21% oxygen, and 1% argon, carbon dioxide, water vapor, and other trace gases. Oxygen is a highly reactive gas whose level is maintained by biological activity in accord with the Gaia Hypothesis.

The *Athenian Hypothesis* states that the climate remains substantially fixed unless externally forced. The external forcing might be volcanoes (whose effects may last seasons to years) or variations in solar radiative output (whose effects may last years to millennia).

Carbon dioxide, a gas containing one carbon atom and two oxygen atoms, is formed by combustion of fossil fuels and by the breathing of man and animals. Photosynthesis converts carbon dioxide back into oxygen and carbon. The oceans remove some carbon dioxide from the atmosphere. Carbon dioxide is mostly transparent to solar radiation, but strongly absorbs thermal or infrared radiation. It may therefore cause a greenhouse effect on Earth, hence it is called a "greenhouse gas."

The *Earth radiation budget* is the balance between absorbed solar radiation and emitted thermal radiation as a function of latitude, longitude, and time, or summarized by hemispheric, global, and annual means.

The *Gaia Hypothesis* suggests that the Earth's biosphere actively controls climate, optimizing conditions for living organisms.

A *General Circulation Model* (or GCM) is a model of the Earth's atmosphere first developed for weather predictions and later modified to predict climatic change.

The *greenhouse effect* is caused by the blanketing effect of such gases as carbon dioxide, which are opaque to thermal radiation, and whose accumulation may warm the Earth. Much of the incident solar radiation penetrating Earth's atmosphere is absorbed. Greenhouse gases such as water vapor, carbon dioxide, ozone, Freon, and so forth absorb the longer wavelengths and thereby trap the reemitted thermal radiation. This secondary absorption heats the Earth more than if the greenhouse gases were absent.

Ice ages were periods of extreme cold when glaciers advanced and covered much of North America and Europe. The last ice age peaked about 18,000 years ago and ended about 8,000 years later. Ice ages are a recent phenomenon, with about 20 occurring during the last 2 million years. They may be explained by the Milankovitch Hypothesis (see below).

A *lapse rate* for the atmosphere is the variation in temperature as a function of altitude.

The *Milankovitch Hypothesis* suggests that Earth's climate varies according to its orbital elements such as precession, obliquity, or eccentricity. This theory helps to explain the timing of the ice ages.

Millibars (mb) are a unit of pressure. At the Earth's surface, the mean pressure equals 1013.25 millibars.

Ozone is a gas composed of three oxygen atoms. The stratospheric ozone layer, which typically extends from 200 to 1 millibar, absorbs much of the incoming solar ultraviolet radiation.

The 18.6-year *Saros lunar cycle* is the period during which the moon-rise and moon-set locations move north and south due to orbital element variations. This cycle has been known for at least 5,000 years, and studies have connected it to the arrangement of the monoliths at Stonehenge.

The *stratosphere* is the region above the troposphere. Stratospheric temperatures rise with increasing altitude. It contains much of the ozone.

The *tropopause* is the boundary between the troposphere and stratosphere.

The *troposphere* is the lower portion of the atmosphere with Hadley cell circulation and vertical convection. The troposphere extends 15 km (about 10 miles) above the Earth's surface at the equator, but may reach only 8 to 10 km at the poles.

Mathematical Terms

Correlation expresses the degree of relationship between two variables. Correlation values vary between $+1$ and -1, where $+1$ shows perfect agreement and -1 shows the exact opposite. A zero value indicates there is no relationship. A significant correlation indicates that a cause-and-effect relationship may exist, but does not prove it. *Autocorrelation* is a measure of the correlation of a variable with measurements of the same variable made at different times.

Fast Fourier Transform (FFT) or *Fourier analysis,* is a standard technique for determining whether time series have periodic behavior. The output is generally a power spectrum.

Independent variables often have zero or low correlations. Two measurements of the same quantity are independent if the data are random or stochastic. Statistical studies assume that measurements of the same quantity are independent. If this assumption of independence breaks down, through persistencies or periodicities, the statistical significance of the conclusions may be overstated.

The *maximum entropy spectral analysis (MESA)* is a method employed to analyze time series to extract the maximum available information using autoregressive techniques. Compared with the FFT, MESA can more easily detect signals in noisy data, allowing successful use of shorter time series.

The *mean* or *average* is the sum of all the values divided by the number of values.

A *linear regression* is the best-fit line through the points representing the dependent and independent variables. Best fit means that the root-mean-square difference between the measured points and the model line are minimized.

The *standard deviation* relates to the variation of the data or the degree of uncertainty in any measurement.

Statistical significance measures how well two independent variables are interrelated. It varies from 0% (no significance) to 100% (complete agreement). If the time series for two variables have nonzero autocorrelations, the statistical significance will be overestimated unless reduced by a proper corrective factor.

The *variance* is the square of the correlation or the percentage of the variability in an effect that can be attributed to a specific cause. For example, a 50% variance means half the changes may be attributed to the variable upon which the correlation, and hence the variance, was calculated. The remaining half may be attributed to other causes, chaotic variations, or measurement errors. A 50% variance has a correlation of .707.

Appendix 2. Solar and Terrestrial Data

Solar Physics Parameters

Solar mass	$1.989 \ 10^{30}$ kg
Solar mass (Earth = 1)	332,830
Equatorial radius	695,000 km
Equatorial radius (Earth = 1)	108.97
Mean density	1.409 g/cm^3
Equatorial rotational period (sidereal)	25 days
Polar rotational period (sidereal)	36 days
Luminosity	$3.827 \ 10^{26}$ watts
Solar irradiance at Earth	1367 W/m^2
Mean surface temperature	5780 °K
Astronomical unit	1.50×10^{11} m
Age	4.7 billion years

Principal chemistry

Hydrogen	92.1%
Helium	7.8%
Oxygen	0.061%
Carbon	0.030%
Nitrogen	0.0084%
Neon	0.0076%
Iron	0.0037%
Silicon	0.0031%
Magnesium	0.0024%
Sulfur	0.0015%
All others	0.0015%

Solar Features that Show an 11-Year Cycle

Feature	Conditions at Sunspot Maximum
Sunspots	Maximum number and area
Faculae	Maximum area
Polar faculae	Minimum area
Plages	Maximum area
Flares	Maximum number
Prominences	Maximum number
Filaments	Maximum number
Chromosphere	Hα reversal has maximum height
Corona	Maximum extent
X-rays	More intense
Lyα line	Brightness is a maximum
Radio emissions	Roughly parallels sunspot areas
Aurorae	Peak 2–3 years after sunspot peak

Some Characteristics of Sunspots

Individual Sunspots

Umbra effective temperature	4240 °K
Umbra to photosphere brightness	0.24
Penumbra effective temperature	5680 °K
Penumbra to photosphere brightness	0.77
Umbra to penumbra area	0.2 (variable)

Sunspot Groups

Individual spots per group	7 to 11
Mean lifetime of groups	6 days
Maximum observed lifetime of a group	134 days
Area of a very large sunspot group	2,256 millionths of solar disk, or 2.6 billion square miles

Sunspot Cycle

Mean sunspot period	11.04 years
Shortest sunspot period	8 years
Longest sunspot period	17 years
Maximum yearly Wolf Sunspot Number	201.3 in 1957

Terrestrial Data

Earth mass	5.975×10^{24} kg
Mean radius	6371 km
Volume	1.08×10^{12} km^3
Surface area	5.10×10^8 km^2
Land area (30%)	1.49×10^8 km^2
Ocean area (70%)	3.61×10^8 km^2
Mean land elevation	860 m
Mean ocean depth	3900 m
Mean sea-level surface pressure	1013.25 mb
Mean surface-air temperature	288°K
Mean atmospheric temperature	255 °K
Mean lapse rate	6.5 °K/km
Mean planetary albedo	30%
Mean absorbed solar radiation	239.4 W/m^2
Mean reemitted thermal radiation	239.4 W/m^2

Dry atmospheric composition by volume:

Nitrogen	78.04%
Oxygen	20.95%
Argon	0.934%
Carbon dioxide	0.0355%
Neon	0.0018%
Helium	0.00052%
Methane	0.00018%
Krypton	0.00011%
Sulfur dioxide	0.00010%
Carbon monoxide	0.00005%
Hydrogen	0.00005%
Others	0.00005%

Solar Activity

Year	R_z	Year	R_z	Year	R_z	Year	R_z	Year	R_z	Year	R_z
1700	5.0	1750	83.4	1800	14.5	1850	66.6	1900	9.5	1950	83.9
1701	11.0	1751	47.7	1801	34.0	1851	64.5	1901	2.7	1951	69.4
1702	16.0	1752	47.8	1802	45.0	1852	54.1	1902	5.0	1952	31.5
1703	23.0	1753	30.7	1803	43.1	1853	39.0	1903	24.4	1953	13.9
1704	36.0	1754	12.2	1804	47.5	1854	20.6	1904	42.0	1954	4.4
1705	58.0	1755	9.6	1805	42.2	1855	6.7	1905	63.5	1955	38.0
1706	29.0	1756	10.2	1806	28.1	1856	4.3	1906	53.8	1956	141.7
1707	20.0	1757	32.4	1807	10.1	1857	22.7	1907	62.0	1957	190.2
1708	10.0	1758	47.6	1808	8.1	1858	54.8	1908	48.5	1958	184.8
1709	8.0	1759	54.0	1809	2.5	1859	93.8	1909	43.9	1959	159.0
1710	3.0	1760	62.9	1810	0.0	1860	95.8	1910	18.6	1960	112.3
1711	0.0	1761	85.9	1811	1.4	1861	77.2	1911	5.7	1961	53.9
1712	0.0	1762	61.2	1812	5.0	1862	59.1	1912	3.6	1962	37.6
1713	2.0	1763	45.1	1813	12.2	1863	44.0	1913	1.4	1963	27.9
1714	11.0	1764	36.4	1814	13.9	1864	47.0	1914	9.6	1964	10.2
1715	27.0	1765	20.9	1815	35.4	1865	30.5	1915	47.4	1965	15.1
1716	47.0	1766	11.4	1816	45.8	1866	16.3	1916	57.1	1966	47.0
1717	63.0	1767	37.8	1817	41.1	1867	7.3	1917	103.9	1967	93.8
1718	60.0	1768	69.8	1818	30.1	1868	37.6	1918	80.6	1968	105.9
1719	39.0	1769	106.1	1819	23.9	1869	74.0	1919	63.6	1969	105.5
1720	28.0	1770	100.8	1820	15.6	1870	139.0	1920	37.6	1970	104.5
1721	26.0	1771	81.6	1821	6.6	1871	111.2	1921	26.1	1971	66.6
1722	22.0	1772	66.5	1822	4.0	1872	101.6	1922	14.2	1972	68.9
1723	11.0	1773	34.8	1823	1.8	1873	66.2	1923	5.8	1973	38.0
1724	21.0	1774	30.6	1824	8.5	1874	44.7	1924	16.7	1974	34.5
1725	40.0	1775	7.0	1825	16.6	1875	17.0	1925	44.3	1975	15.5
1726	78.0	1776	19.8	1826	36.3	1876	11.3	1926	63.9	1976	12.6
1727	122.0	1777	92.5	1827	49.6	1877	12.4	1927	69.0	1977	27.5
1728	103.0	1778	154.4	1828	64.2	1878	3.4	1928	77.8	1978	92.5
1729	73.0	1779	125.9	1829	67.0	1879	6.0	1929	64.9	1979	155.4
1730	47.0	1780	84.8	1830	70.9	1880	32.3	1930	35.7	1980	154.6
1731	35.0	1781	68.1	1831	47.8	1881	54.3	1931	21.2	1981	140.4
1732	11.0	1782	38.5	1832	27.5	1882	59.7	1932	11.1	1982	115.9
1733	5.0	1783	22.8	1833	8.5	1883	63.7	1933	5.7	1983	66.6
1734	16.0	1784	10.2	1834	13.2	1884	63.5	1934	8.7	1984	45.9
1735	34.0	1785	24.1	1835	56.9	1885	52.2	1935	36.1	1985	17.9
1736	70.0	1786	82.9	1836	121.5	1886	25.4	1936	79.7	1986	13.4
1737	81.0	1787	132.0	1837	138.3	1887	13.1	1937	114.4	1987	29.2
1738	111.0	1788	130.9	1838	103.2	1888	6.8	1938	109.6	1988	100.2
1739	101.0	1789	118.1	1839	85.7	1889	6.3	1939	88.8	1989	157.7
1740	73.0	1790	89.9	1840	64.6	1890	7.1	1940	67.8	1990	141.8
1741	40.0	1791	66.6	1841	36.7	1891	35.6	1941	47.5	1991	145.2
1742	20.0	1792	60.0	1842	24.2	1892	73.0	1942	30.6	1992	94.4
1743	16.0	1793	46.9	1843	10.7	1893	85.1	1943	16.3	1993	54.6
1744	5.0	1794	41.0	1844	15.0	1894	78.0	1944	9.6	1994	29.9
1745	11.0	1795	21.3	1845	40.1	1895	64.0	1945	33.2	1995	19.1
1746	22.0	1796	16.0	1846	61.5	1896	41.8	1946	92.6	1996	
1747	40.0	1797	6.4	1847	98.5	1897	26.2	1947	151.6	1997	
1748	60.0	1798	4.1	1848	124.7	1898	26.7	1948	136.3	1998	
1749	80.9	1799	6.8	1849	96.3	1899	12.1	1949	134.7	1999	

This table lists the Wolf or Zurich Sunspot Numbers (R_z) for each year from 1700 to 1988. These yearly means are the most commonly used solar index in sun/climate studies. For the years 1645 to 1699 (the Maunder Minimum), the sunspot numbers are close to zero (see Figure 2.13).

The Solar Spectral Irradiance

λ	$I(\lambda)$	λ	$I(\lambda)$	λ	$I(\lambda)$
0.250	69.3	0.656	1524.0	1.477	307.3
0.255	86.0	0.668	1531.0	1.497	300.4
0.260	114.1	0.690	1420.0	1.520	292.8
0.265	160.0	0.710	1399.0	1.539	275.5
0.270	179.0	0.718	1374.0	1.558	272.1
0.275	177.0	0.724	1373.0	1.578	259.3
0.280	210.4	0.740	1298.0	1.592	246.9
0.285	273.0	0.753	1269.0	1.610	244.0
0.290	363.8	0.758	1245.0	1.630	243.5
0.295	505.0	0.763	1223.0	1.646	234.8
0.300	535.9	0.768	1205.0	1.678	220.5
0.305	558.3	0.780	1183.0	1.740	190.8
0.310	622.0	0.800	1148.0	1.800	171.1
0.315	692.7	0.816	1091.0	1.860	144.5
0.320	715.1	0.824	1062.0	1.920	135.7
0.325	832.9	0.832	1038.0	1.960	123.0
0.330	961.9	0.840	1022.0	1.985	123.9
0.335	931.9	0.860	998.7	2.005	113.0
0.340	900.6	0.880	947.2	2.035	108.5
0.345	911.3	0.905	893.2	2.065	97.5
0.350	975.5	0.915	868.2	2.100	92.4
0.360	975.9	0.925	829.7	2.148	82.4
0.370	1119.9	0.930	830.3	2.198	74.6
0.380	1103.8	0.937	814.0	2.270	68.3
0.390	1033.8	0.948	786.9	2.360	63.8
0.400	1479.1	0.965	768.3	2.450	49.5
0.410	1701.3	0.980	767.0	2.500	48.5
0.420	1740.4	0.994	757.6	2.600	38.6
0.430	1587.2	1.040	688.1	2.700	36.6
0.440	1837.0	1.070	640.7	2.800	32.0
0.450	2005.0	1.100	606.2	2.900	28.1
0.460	2043.0	1.120	585.9	3.000	24.8
0.470	1987.0	1.130	570.2	3.100	22.1
0.480	2027.0	1.145	564.1	3.200	19.6
0.490	1896.0	1.161	544.2	3.300	17.5
0.500	1909.0	1.170	533.4	3.400	15.7
0.510	1927.0	1.200	501.6	3.500	14.1
0.520	1831.0	1.240	477.5	3.600	12.7
0.530	1891.0	1.270	442.7	3.700	11.5
0.540	1898.0	1.290	440.0	3.800	10.4
0.550	1892.0	1.320	416.8	3.900	9.5
0.570	1840.0	1.350	391.4	4.000	8.6
0.593	1768.0	1.395	358.9		
0.610	1728.0	1.443	327.5		
0.630	1658.0	1.463	317.5		

The solar spectral irradiance $I(\lambda)$ as a function of wavelength (λ) in microns at 1 astronomical unit. The spectral irradiance units are W/m^2 per micron. The data are derived from Neckel and Labs (1984) and include about 99% of the total irradiance (see Figure 3.3).

Appendix 3. A Technical Discussion of the Statistical Basis of Sun/Weather/ Climate Studies

Bayesian Estimation

Bayesian estimation allows new data to be added a posteriori to existing effects (a priori ideas) with a view toward understanding how the new data alter the previous determination, theory, or idea. Many sun/weather effects disappear with surprising frequency following the publication of new data. As recent chaos theories suggest, weather and climate are inherently variable, and hence it is not surprising that if a physical effect on sun/weather/climate were correct, it could be masked during certain intervals by the inherent variability of the weather and climate. Thus, new data should not just be said "to disagree" with a particular effect, but rather a new estimation of reliance based on Bayesian statistics should provide a new determination of the effect's reliance.

Given, for example, a variable with a predicted or calculated mean, μ_0, and standard deviation, σ_0, and the new information of n independent points of additional data with mean, x, and standard deviation, σ, the new mean and standard deviations are:

$$\mu_1 = \frac{nx\sigma_0^2 + \mu_0\sigma^2}{n\sigma_0^2 + \sigma^2} \tag{A-1}$$

and

$$\sigma_1 = \sqrt{\frac{\sigma^2\sigma_0^2}{n\sigma_0^2 + \sigma^2}} \tag{A-2}$$

If the new data have few new points (n small), even if the new mean, x, is zero, μ_1 is not significantly changed, nor is σ_0. Equations A-1 and A-2 (or variants of this form to include the number of original points) should be used to see how significantly the new material really alters the statistical inference.

Consideration should be given to the fact that the weather and climate have a degree of inherent variability, which will mask many effects for brief intervals.

Chree Analyses

Often, sun/weather/climate analyses have used superposed epoch (or Chree) analyses to demonstrate an effect. For example, do solar sectors boundaries crossing the Earth trigger some effect? By superposing many such crossings, the validity of the effect can be determined through Chree analyses. In many sun/climate studies, error bars are not often given, or, if they are, it is uncertain how the bars are obtained. This section outlines how the levels of significance may be calculated from these methods.

In a Chree analysis, assume one has N_1 superposed time periods (with N_2 independent data points in each period). If one considers the data of an individual event, yi, to be placed in a row, i, and the time of the data, relative to the epoch y_{ij} placed in the column, j, then the mean and standard deviations of the rows and columns may be computed separately. That is: $\mu_{ij}\sigma_i$ and $\mu_{ij}\sigma_j$. If an "effect" is found, it is expressed as:

$$y_{ij} = \mu_i + \epsilon_{ij} \tag{A-3}$$

for $i=1$, N_1 and $j=1$, N_2, where μ_j is the mean superposed epoch and ϵ_{ij} are the normally distributed errors with a variance σ^2.

To test the null hypothesis (that all the means μ_j are equal), one compares two estimates of σ^2, the variance of the columns and of the rows:

$$\sigma_i^2 = \sum \frac{(y_{ij} - y_i)^2}{N_2 - 1} \tag{A-4}$$

The mean row variance is calculated as:

$$\sigma_i^2 = \sum \frac{\sigma_i^2}{N_1} \tag{A-5}$$

and the mean column variance is calculated as:

$$\sigma_j^2 = \sum \frac{\sigma_j^2}{N_2} \tag{A-6}$$

If the null hypothesis is true, removing μ_{ij} has little effect on the result, and equations (A-5) and (A-6) are approximately equal. Using the F test:

$$F = \frac{\sigma_1^2}{\sigma_j^2} \tag{A-7}$$

is a value of a random variable having the F distribution with $N_2 - 1$ and $N_1 - 1$ degrees of freedom.

Since the sample variance σ_i^2 can be expected to exceed the sample variance σ_j^2 once the mean has been removed, the null hypothesis is rejected if F exceeds $\{F\alpha$—the F distribution to some α level of significance$\}$. For example, removing the mean of a test gives a ratio of F for 10 days of independent data, in a Chree analysis, with 101 superposed epochs, we examine $N_2 = 10$ and $N_1 = 101$, or $F_{9,100}$, which is 1.97 at the 5% level of significance. If F exceeds 1.97 (reducing the variance by half), it is significant at the 5% level; if not, it is less significant. If $F \sim 1$, the results are insignificant.

Cross-Spectral Analysis

Scientists who must deal with two variables often study their behavior individually, without regard to their joint behavior. A common periodicity in the individual power spectra often leads to the belief that the two variables may be causally related. Rather than review this entire subject, see Bendat and Piersol (1971) for the use of cross-spectral analyses to test any related "common" periodicities. Cross spectra will provide the power and phasing between the two sets of data at various frequencies. Even if both spectra have common periodicities, unless they are in phase (or have a constant phase shift), no positive result will emerge; hence, this technique is superior to straightforward individual power spectral analyses.

Independence of Data

Let us now consider a problem that often enters the work of many researchers in the field of sun/weather/climate. Statistical analysis methods assume data independence. Geophysical phenomena often do not fulfill this requirement because they are correlated in space or time. Two main causes of this lack of independence are (1) persistency and common periodicities in the solar and meteorological records, creating data which are not strictly independent and thus have fewer statistically-independent real samples than are apparently present, and (2) the possibility that a real physical effect exists and some solar influences are affecting the meteorological records. Let us disregard the possible real physical influences and consider how the statistical significance is often overestimated. For further discussions, see Volland and Schafer (1979) and Schatten et al. (1979).

An example will clarify this effect. Planetary waves are large-scale variations in surface pressure that typically persist for 5 days. Four planetary waves typically exist at any given time. Solar activity influences the interplanetary medium (such as sector boundaries or high-speed streams) in a fraction of the 27-day solar rotation period because an active region takes only a few days to cross the central meridian. Thus, solar activity also has periodicities on the order of a few days. Additionally, both the Earth's weather and the sun's "weather" remain roughly the same from one day to the next, as does the solar appearance. On longer time scales, the sun has an 11-year cycle and the 80- to 90-year Gleissberg cycle (which may be more of a persistency than a strict periodicity). Similarly, the Earth's climate may have a natural persistency due to some phenomena requiring time scales of a few years to dissipate (for example, oceanic heat transport, volcanic aerosol effects, etc.), which would provide a climatic persistency over certain long time scales.

Such periodicities mean each point in any particular data set is not strictly independent from the next. Most statistical analyses would convert a 1 SD effect into a many SD effect simply due to the lack of data independence, causing the researcher to find a false effect purely by chance. To eliminate this possibility, one must consider and remove the common periodicity and persistency effects without eliminating any possible real effect.

Bibliography

The first part of the bibliography lists *major* papers mentioned in the text of the chapters. Many early references and those dealing with non-sun/climate topics are considered to be minor and are not included in the bibliography. A few background references not explicitly mentioned in the text are also included in the chapter bibliography. The second part provides a fairly complete chronological listing of sun/climate-related papers and books to 1900. After 1900, we list papers that, in our judgment, appear better than average or important. The final section lists books primarily devoted to sun/weather/climate relationships.

References Cited in the Text

Chapter 1. Introduction

Herschel, W., 1796. Some remarks on the stability of the light of the sun. *Philosophical Transactions of the Royal Society, London,* 166.

Herschel, W., 1801. Observations tending to investigate the nature of the sun, in order to find the causes or symptoms of its variable emission of light and heat. *Philosophical Transactions of the Royal Society, London,* 265.

Herschel, W., 1801. Additional observations tending to investigate the symptoms of its variable emission of light and heat of the sun, with trials to set aside darkening glasses, by transmitting the solar rays through liquids and a few remarks to remove objections that might be made against some of the arguments contained in the former paper. *Philosophical Transactions of the Royal Society, London,* 354.

Meldrum, C., 1872. On a periodicity in the frequency of cyclones in the Indian Ocean south of the equator. *Nature, 6,* 357–358.

Meldrum, C., 1872. On a periodicity in the frequency of cyclones in the Indian Ocean south of the equator. *British Association Report,* 56–58.

Meldrum, C., 1885. On a supposed periodicity of the cyclones of the Indian Ocean south of the equator. *British Association Report,* 925–926.

Chapter 2. Observations of the Sun

Babcock, H. W., 1961. The topology of the sun's magnetic field and the twenty-two-year cycle. *Astrophysical Journal, 133,* 572–587.

Brown, G. M. and W. R. Williams, 1969. Some properties of the day-to-day variability of So(H). *Planetary and Space Sciences, 17,* 455–470.

Eddy, J. A., 1976. The Maunder minimum. *Science, 192,* 1189–1192.

Fraunhofer, J., 1817. Bestimmung des Brechungs- und Farbenzerstreuungs- Vermogens verschiedener Glasarten, in Bezug auf die Vervollkommung achromatischer Fern-rorhe. *Denkschriften der königlichen Akademie der Wissenschaften zu München: Classe der Mathematik und Naturwissenschaften, 5,* 193–226.

Friedman, H., 1986. *Sun and Earth.* Scientific American Books, Inc. New York, 251 pp.

Gautier, A., 1844. Recherches relatives a l'influence que le nombre et la permanence des taches observees sur le disque du soliel peuvent execer sur les temperatures terrestres. *Annales de Chemie et de Physique, 3,* 12.

Hale, G. E., 1908. On the probable existence of a magnetic field in sun-spots. *Astrophysical Journal, 28,* 315–343.

Herschel, W., 1801. Observations tending to investigate the nature of the sun, in order to find the causes or symptoms of its variable emission of light and heat. *Philosophical Transactions of the Royal Society, London,* 265.

Herschel, W., 1801. Additional observations tending to investigate the symptoms of its variable emission of light and heat of the sun, with trials to set aside darkening glasses, by transmitting the solar rays through liquids and a few remarks to remove objections that might be made against some of the arguments contained in the former paper. *Philosophical Transactions of the Royal Society, London,* 354.

Humboldt, A. von, 1850. *Cosmos: A Sketch of a Physical Description of the Universe.* 5 vols., New York.

Luby, W. A., 1942. The mean period of the sunspots. *Popular Astronomy, 50,* 537–588.

Maunder, A. S. D. and E. W. Maunder, 1908. *The Heavens and Their Story.* R. Culley, London, 357 pp.

Maunder, E. E., 1922. The prolonged sunspot minimum, 1645–1715. *Journal of the British Astronomical Society, 32,* 140.

Maunder, E. W., 1894. A prolonged sunspot minimum. *Knowledge, 17,* 173–176.

Mitchell, W. M., 1915. *Popular Astronomy, 23,* 94, 155–156.

Newton, H. W. and P. Leigh-Smith, 1941. Anomalous behaviour of the 11-year solar cycle in the seventeenth century. *Journal of the British Astronomical Association, 51,* 101–102.

Nicholson, S. B., 1933. *The Solar Cycle.* Leaflet 50. Astronomical Society of the Pacific.

Noyes, R. W., 1982. *The Sun. Our Star.* Harvard University Press, Cambridge, 263 pp.

Ohl, A. I. and G. I. Ohl, 1979. A new method of very long-term prediction of solar activity. *Proceedings of the Conference on Solar-Terrestrial Predictions* (R. Donnelly, ed.), V, ii, 258.

Schatten, K. H., P. H. Scherrer, L. Svalgaard, and J. M. Wilcox, 1978. Using dynamo theory to predict the sunspot number during solar cycle 21. *Geophysical Research Letters, 5,* 411.

Schatten, K. H. and S. Sofia, 1987. Forecast of an exceptionally large even-numbered solar cycle. *Geophysical Research Letters, 14,* 632–635.

Schwabe, A. N., 1844. Sonnen-Beobachtungen im Jahre 1843. *Astronomische Nachrichten, 21,* 233.

Silverman, S. M., 1992. Secular variation of the aurora for the past 500 years. *Reviews of Geophysics and Space Physics, 30,* 333–351.

Spoerer, G., 1889. Über die Periodicität de Sonnenflecken seit dem Jahre 1618. *Nova Acta der Ksl. Leop.-Carol. Deutschen Akademie der Naturforscher, 53,* 283–324.

Stetson, H. T., 1947. *Sunspots and Their Effects. From a Human Point of View.* McGraw-Hill, New York, 201 pp.

Thiele, T. N., 1859. De macularum solis antiquioribus quibusdam observationibus Hafniae Institutis. *Astronomische Nachrichten, 50,* 257–259.

Virgil, 70–19 B.C. *Georgics.*

Wilson, A., 1774. Observations of the solar spots. *Philosophical Transactions of the Royal Society, London, 64,* 1–30.

Wolf, R., 1868. In *Astronomische Mittheilungen.*

Chapter 3. Variations in Solar Brightness

Abbot, C. G., 1910. The solar constant of radiation. *Smithsonian Institution Annual Report,* 319.

Abbot, C. G., 1934. *The Sun and the Welfare of Man.* New York, Smithsonian Institution Series.

Abbot, C. G., 1958. *Adventures in the World of Science.* Public Affairs Press, Washington, D.C., 150 pp.

Baliunas, S. L. and A. H. Vaughan, 1985. Stellar activity cycles, *Annual Reviews of Astronomy and Astrophysics, 23,* 379–412.

Clayton, H. H., 1932. Solar variations and atmospheric pressure. *Science, 77,* 568.

Eyer, L., M. Grenon, J.-L. Falin, M. Froeschle, and F. Mignard, 1994. Variable stars with the Hipparchos satellite. In *The Sun as a Variable Star: Solar and Stellar Irradiance Variations* (J. Pap et al., eds.). Kluwer Academic Publishers, Dordrecht, pp. 91–96.

Foukal, P. and J. Lean, 1990. An empirical model of total solar irradiance variation between 1874 and 1988. *Science, 247,* 556–558.

Foukal, P. V., P. E. Mack, and J. E. Vernazza, 1977. The effect of sunspots and faculae on the solar constant. *Astrophysical Journal, 215,* 952.

Hickey, J. R., L. L. Stowe, H. Jacobowitz, P. Pellegrino, R. H. Maschhoff, F. House, and T. H. VonderHaar, 1980. Initial solar irradiance determinations from Nimbus 7 cavity radiometer measurements. *Science, 208,* 281–283.

Herschel, W., 1801. Observations tending to investigate the nature of the sun, in order to find the causes or symptoms of its variable emission of light and heat. *Philosophical Transactions of the Royal Society, London,* 265.

Herschel, W., 1801. Additional observations tending to investigate the symptoms of its variable emission of light and heat of the sun, with trials to set aside darkening glasses, by transmitting the solar rays through liquids and a few remarks to remove objections that might be made against some of the arguments contained in the former paper. *Philosophical Transactions of the Royal Society, London,* 354.

Herschel, W., 1807. Letter to Bode on the variation of solar heat. *Bode's Jahrbuch,* 242.

Hoyt, D. V., H. L. Kyle, J. R. Hickey, and R. H. Maschhoff, 1992. The Nimbus-7 solar total irradiance: A new algorithm for its derivation. *Journal of Geophysical Research, 97A* 51–63.

Humphreys, W. J., 1910. Solar disturbances and terrestrial temperatures. *Astrophysical Journal, 32,* 97–111.

Kondratyev, K. Ya. and G.A. Nikolsky, 1970. Solar radiation and solar activity. *Journal of the Royal Meteorological Society, 96,* 509.

Langley, S. P., 1876. Measurement of the direct effect of sun-spots on terrestrial climates. *Monthly Notices of the Royal Astronomical Society, 37,* 5–11.

Langley, S. P., 1884. *Researches on the solar heat and its absorption by the Earth's atmosphere. A report of the Mount Whitney Expedition.* Professional Paper 15, Signal Service, Washington, D.C.

Lean, J., 1989. Contribution of ultraviolet irradiance variations to changes in the sun's total irradiance. *Science, 244,* 197–200.

Lean, J., 1991. Variations in the Sun's radiative output. *Reviews of Geophysics and Space Physics, 29,* 505–535.

Livingston, W. C., 1990. Secular change in equivalent width of C5380, 1978–90. In *Climate Impact of Solar Variability* (K. H. Schatten and A. Arking, eds.). NASA Conference Publication CP-3086, 336–340.

Lockwood, G. W. and B. A. Skiff, 1990. Some insights on solar variability from precision stellar astronomical photometry. In *Conference on the Climate Impact of Solar Variability* (K. H. Schatten and A. Arking, eds.). NASA Conference Publication CP-3086, 8–15

Lockwood, G. W., B. A. Skiff, S. L. Baliunas, and R. R. Radick, 1992. Long-term solar brightness changes estimated from a survey of Sun-like stars. *Nature, 360,* 653–655.

Lockyer, J. N., 1895. Observations of sunspot spectra. I. The widening of iron lines and unknown lines in relation to the sunspot period. *Nature, 51,* 448–449.

Schatten, K. H., 1993. Heliographic latitude dependence of the Sun's irradiance. *Journal of Geophysical Research, 98,* 18907–18910.

Secchi, A., 1877. *Le Soleil,* Vols. 1–2. Paris: Gurthier-Villars.

Smyth, C. P., 1856. Note on the constancy of solar radiation. *Royal Astronomical Society, Memoirs, 25,* 125–131.

Sterne, T. E. and N. Dieter, 1958. The constancy of the solar constant. *Smithsonian Contributions to Astrophysics, 3,* 9–12.

Willson, R. C., C. H. Duncan, and J. Geist, 1980. Direct measurement of solar luminosity variation. *Science, 207,* 177–179.

Willson, R. C., S. Gulkis, M. Janssen, H. S. Hudson, and G. A. Chapman, 1981. Observation of solar irradiance variability. *Science, 211,* 700–702.

Zhang, Q., W. H. Soon, S. L. Baliunas, G. W. Lockwood, B. A. Skiff, and R. R. Radick, 1994. A method of determining possible brightness variations of the sun in past centuries from observations of solar-type stars. *Astrophysical Journal Letters, 427,* L111–114.

Chapter 4. Climate Measurement and Modeling

Agee, E. N., 1982. A diagnosis of twentieth-century temperature records at West Lafayette, Indiana. *Climatic Change, 4,* 399–418.

Borucki, W. J., J. B. Pollack, O. B. Toon, H. T. Woodward, and D. R. Wiedman, 1980. The influence of UV variations on climate. *Proceedings of the Conference on the Ancient Sun* (R. O. Pepin, J. A. Eddy, and R. B. Merrill eds.). Pergamon, New York, 513–522.

Chandra, S., 1991. The solar UV related changes in total ozone from a solar rotation to a solar cycle. *Geophysical Research Letters, 18,* 837.

Crowley, T. J. and M. K. Howard, 1990. Testing the sun-climate connection with paleoclimate data. In *Conference Proceedings, Climate Impact of Solar Variability* (K. H. Schatten and A. Arking, eds.). NASA Conference Publication CP-3086, 81–89.

Crowley, T. J. and R. J. North, 1991. *Paleoclimatology*. Oxford University Press, New York.

Damon, P. E. and O. R. White (eds.), 1977. *The Solar Output and Its Variations*. Colorado Associated University Press, Boulder, 429–445.

Eddy, J. A., 1977. Climate and the changing sun. *Climatic Change, 1,* 173–190.

Engle, H. W. and C. W. Groetsch, 1987. *Inverse and Ill-posed Problems*. Academic Press.

Friis-Christensen, E. and K. Lassen, 1991. Length of the solar cycle: An indicator of solar activity closely associated with climate. *Science, 254,* 698–700.

Gunst, R. F., B. Sabyasachi, and R. Brunell, 1992. Defining and estimating global mean temperature anomalies. *Journal of Climate, 6,* 1368–1374.

Hansen, J., A. Lacis, D. Rind, G. Russell, P. Stone, I. Fung, R. Ruedy, and J. Lerner, 1984. Climate sensitivity: Analysis of feedback mechanisms. In *Climate Processes and Climate Sensitivity*. Geophysical Monograph 29, American Geophysical Union, 130–163.

Hansen, J. E., 1993. Proposed Climsat mission. In *Long-term monitoring of global climate forcings and feedbacks* (Hansen, J. E. et al., eds), NASA-CP-3234, 101 pp.

Heath, D. F. and M. P. Thekaekara, 1977. The solar spectrum between 1200 and 3000 A.D. In *The Solar Output and its Variations*. Colorado Associated University Press, Boulder, 193–212.

Kelly, P. M. and T. M. L. Wigley, 1992. Solar cycle length, greenhouse forcing, and global climate. *Nature, 360,* 328–330.

Kippenhahn, R. and A. Weigert, 1990. *Stellar Structure and Evolution*. Springer-Verlag, Berlin.

Lacis, A. A. and B. E. Carlson, 1992. Keeping the sun in proportion. *Nature, 360,* 297.

Manabe, S. and R. F. Strickler, 1964. Thermal equilibrium of the atmosphere with convective adjustment. *Journal of the Atmospheric Sciences, 21,* 361–385.

McLaughlin, D. W. (ed.), 1984. *Inverse Problems*. American Mathematical Society, Providence, Rhode Island, *14,* 189 pp.

Mitchell, J. M., Jr., 1953. On the causes of instrumentally observed secular temperature trends. *Journal of Meteorology, 10,* 244–261.

Piexoto, J. P. and A. H. Oort, 1992. *Physics of Climate*. American Institute of Physics, New York, 520 pp.

Ramanathan, V. and J. A. Coakley, 1978. Climate modeling through radiative-convective models. *Reviews of Geophysics and Space Physics, 16,* 465–489.

Schatten, K. H., 1995. An Atmospheric Radiative-Convective Model: Solar Forcings. *Eos, Transactions, American Geophysical Union,* S239, SH52-A2.

Schlesinger, M. E. and N. Ramankutty, 1992. Implications for global warming of intercycle solar irradiance variations. *Nature, 360,* 330–333.

Stolarski, R.S., P. Bloomfield, R. MacPeters, and J. Herman, 1991. Total ozone trends deduced from Nimbus 7 TOMS data. *Geophysical Research Letters, 18,* 1015.

Wigley, T. M. L. and S. C. B. Raper, 1990. Climatic change due to solar irradiance changes. *Geophysical Research Letters, 17,* 2169–2172.

Chapter 5. Temperature

Angot, A., 1903. On the simultaneous variations of sun spots and of terrestrial atmospheric temperatures. *Monthly Weather Review, 31,* 371. (Translated from *Annuaire de la Société Météorologique de France,* Juin 1903)

Arctowski, H., 1909. L'enchainement des variations climatiques. *Société Belge d'Astronomie,* Brussells.

Arrhenius, S., 1896. On the influence of carbonic acid in the air upon the temperature of the ground, *Philosophical Magazine 41,* 237.

Blanford, H. F., 1891. The paradox of the sunspot cycle in meteorology. *Nature, 43,* 583–587.

Bruckner, E., 1890. *Klimaschwankungen seit 1700 nebst Bemerkungen über die Klimaschungen der Diluviatzeit, Geographische Abhandlungen, 4,* no. 2. Vienna.

Brunt, D., 1925. Periodicities in European weather. *Philosophical Transactions of the Royal Society, London, 225A,* 247.

Clayton, H. H., 1940. The 11-year and 27-day solar periods in meteorology. *Smithsonian Miscellaneous Collections, 39,* No. 5, 20 pp.

Clayton, H. H., 1943. *Solar Relations to Weather.* Clayton Weather Service, Canton, Mass., Vol 2, 439 pp.

Craig, R. A., 1965. In *Proceedings of a Seminar on Possible Responses of Weather Phenomena to Variable Extraterrestrial Influences.* NCAR Technical Note TN-8.

Currie, R. G., 1974. Solar cycle signal in surface air temperature, *Journal of Geophysical Research, 79,* 5657–5660.

Currie, R. G., 1979. Distribution of solar cycle signal in surface air temperature over North America. *Journal of Geophysical Research, 84,* 753–761.

Currie, R. G., 1981a, Solar cycle signal in air temperature in North America: Amplitude, gradient, phase and distribution. *Journal of the Atmospheric Sciences, 38,* 808–818.

Currie, R. G., 1981b, Evidence for 18.6-year M_N signal in temperature and drought conditions in North America since 1800 AD. *Journal of Geophysical Research, 86,* 11055–11064.

Currie, R. G., 1993. Luni-solar 18.6 and solar cycle 10–11 year signals in USA air temperature records. *International Journal of Climatology, 13,* 31–50.

Deshara, M. and K. Cehak, 1970. A global survey on periodicities in annual mean temperatures and precipitation totals. *Archive für Meteorologie, Geophysik und Bioklimatologie, Ser. B, 18,* 253–268.

Easton, 1905. Zur Periodizität der solaren und klimatischen Schwankungen. *Petermanns Geographische Mitteilungen.*

Flaugergues, 1823. *Zach Correspondance Astronomique, 9.*

Fritz, H., 1893. Die perioden solarer und terrestrischer Erscheinungen. *Zurich Vierteljahresschriften, 38,* 77–107.

Goad, J., 1686. *Astrometeorologica, or Aphorisms and Discourses on the Bodies Celestial, Their Nature and Influences.* Printed by J. Rawlings for O. Blagrave, London, 509 pp.

Greg, R. P., 1860. On the periodicity of the solar spots, and induced meteorological disturbances. *Philosophical Magazine, 20,* ser. 4, 246–247.

Gruithuisen, Fr.v.P., 1826. Naturwissenschaftlicher Reisebericht. *Kastner's Archiv für die gesammte Naturiehre, 8,* Nürnberg.

Gunther, S., 1899. *Handbuch der Geophysik, 2.* Aufl. II.

Hahn, F. G., 1877. *Über die Beziehungen der Sonnenfleckenperiode zu meteorologischen Erscheinungen.* Leipzig.

Hann, J., 1908. *Handbuch der Klimatologie,* 3 volumes, Engelhorn, Stuttgart.

Helland-Hansen, B. and F. Nansen, 1920. *Temperature Variations in the North Atlantic Ocean and the Atmosphere: Introductory Studies on the Cause of Climatological Variations.* Smithsonian Miscelaneous Collection, Washington, D.C., 70, no. 4, 408 pp.

Herman, J. R. and R. A. Goldberg, 1978. *Sun, Weather, and Climate.* NASA, Washington, D.C., 360 pp.

Hill, S.A., 1879. Über eine zehnjährige Periode in der jährlichen Änderung der Temperatur und des Luftdruckes in Nord-Indien. *Zeitschrift der Österreichischen Gesellschaft für Meteorologie, 14.*

Humphreys, W. J., 1910. Solar disturbances and terrestrial temperatures. *Astrophysical Journal, 32,* 97–111.

King, J. W., 1973. Solar radiation changes and the weather, *Nature, 245,* 443–446.

King, J. W., 1974. Weather and the earth's magnetic field. *Nature, 247,* 131–134.

King, J. W., E. Hurst, A. J. Slater, P. A. Smith, and B. Tomkin, 1974. Agriculture and sunspots, *Nature, 252,* 2–3.

Kircher, A., 1671. *Iter Exstaticum Coeleste,* Herbipoli.

Koppen, W., 1873. On temperature cycles. *Nature, 9,* 184–185.

Koppen, W., 1914. Lufttemperaturen, Sonnenflecke und Vulkanausbruche. *Meteorologische Zeitung, 31,* 305–328.

Labitze, K. and H. van Loon, 1988. Associations between the 11-year solar cycle, the quasi-biennial oscillation and the atmosphere. Part I. The troposphere and stratosphere in the northern hemisphere in winter. *Journal of Atmospheric and Terrestrial Physics, 50,* 197–206.

Landsberg, H. E., J. M. Mitchell, Jr., and H. L. Crutcher, 1959. Power spectrum analysis of climatological data for Woodstock College, Maryland. *Monthly Weather Review, 87,* 283.

Lawrence, E. N., 1965. Terrestrial climate and the solar cycle. *Weather, 20,* 334–343.

Lee, R. B., 1992. Implications of solar irradiance variability upon long-term changes in the Earth's atmospheric temperatures. *Journal of the National Technology Association, 65,* 65–71.

Mielke, J., 1913. Die Temperaturschwankungen 1870–1910 in ihrem Verhältnis zu der 11-jährigen Sonnenfleckenperiode. *Archiv der Deutschen Seewarte, 36,* No. 3.

Newcomb, S., 1908. A search for fluctuations in the sun's thermal radiation through their influence on terrestrial temperature. *Transactions of the American Philosophical Society, Philadelphia, 21,* 309–387.

Nordmann, C., 1903. Connection between sunspots and atmospheric temperature. *Nature, 68,* 162.

North, G. R., J. G. Mengel, and D. A. Short, 1983. Climatic response to a time varying solar constant. In *Weather and Climate Responses to Solar Variations* (B. M. McCormac, ed.). Colorado Associated University Press, Boulder, 243–255.

Riccioli, J. B., 1651. *Almagestrum Novum,* Bononiae, Italy, 2 vols., 763 pp + 676 pp.

Schuster, A., 1884. Quoted in Blanford, H. F., 1891. The paradox of the sunspot-cycle in meteorology. *Nature, 43,* 583–587.

Shaw, D., 1965. Sunspots and temperatures. *Journal of Geophysical Research, 70,* 4997.

Shaw, Sir N., 1928. *Manual of Meteorology,* Vol. 2., *Comparative Meteorology.* University Press, Cambridge, England.

Smyth, C. P., 1870. On supra-annual cycles of temperature in the earth's surface-crust. *Royal Society Proceedings, 18,* 311–312.

Smyth, C. P., 1870. On supra-annual cycles of temperature in the earth's surface-crust.

Philosophical Magazine, 40, 58–59.

Stone, E. J., 1871. On an approximately decennial variation of the temperature at the Observatory at the Cape of Good Hope between the years 1841 and 1870, viewed in connexion with the variation of the solar spots. *Royal Society Proceedings, 19,* 389–392.

Stone, E. J., 1871. On an approximately decennial variation of the temperature at the Observatory at the Cape of Good Hope between the years 1841 and 1870, viewed in connexion with the variation of the solar spots. *Philosophical Magazine, 42,* 72–75.

Williams, C. J. B., 1883. On the cold in March, and absence of sunspots. *Nature, 28,* 102–103.

Chapter 6. Rainfall

Baur, F., 1949. The double oscillation of the atmospheric circulation in the temperate zone during the sunspot cycle (German). *Meteorologische Rundschau, 2,* 10–15.

Bell, P. R., 1982. Predominant periods in the time series of drought area index for the western high plains AD 1700 to 1962. *Proceedings of the Workshop on Solar Constant Variations,* NASA Conference Paper, Washington, D. C.

Bollinger, C. J., 1945. The 22-year solar pattern of rainfall in Oklahoma and Kansas. *Bulletin of the American Meteorological Society, 26,* 376–383.

Brooks, C. E. P., 1923. Variations in the level of Lake George, Australia. *Nature, 112,* 918–919.

Brooks, C. E. P., 1928. Periodicities in the Nile floods. *Royal Meteorological Society, Memoirs, 2,* 12, 26 pp.

Bruckner, E., 1890. Klimaschwankungen seit 1700 [Climate variations since 1700]. *Geographische Abhandlungen, 14,* 325 pp.

Chambers, F., 1878. Sun and earth. *Nature, 18,* 619–620.

Chambers, F., 1878. Sun-spots and weather. *Nature, 18,* 567–568.

Clayton, H. H., 1923. *World Weather, Including a Discussion of the Influence of Variations of Solar Radiation on the Weather and of the Meteorology of the Sun.* Macmillan, New York, 393 pp.

Clayton, H. H., 1940. The 11-year and 27-day solar periods in meteorology. *Smithsonian Miscellaneous Collections, 39,* no. 5, 20 pp.

Clayton, H. H., 1943. *Solar Relations to Weather.* Clayton Weather Service, Canton, Mass., vol 2, 439 pp.

Clough, H. W., 1920. An approximate 7-year period in terrestrial weather with solar correlation. *Monthly Weather Review, 48,* 593–597.

Currie, R. G., 1987. On bistable phasing of 18.6-year induced drought and flood in the Nile records since AD 650. *Journal of Climatology, 7,* 373–379.

Currie, R. G. and D. P. O'Brien, 1992. Deterministic signals in USA precipitation records. Part II. *International Journal of Climatology, 12,* 182–184.

Dawson, G. M., 1874. The fluctuations of the American lakes and the development of sun-spots. *Nature, 9,* 504.

Deshara, M. and K. Cehak, 1970. A global survey on periodicities in annual mean temperatures and precipitation totals. *Archiv für Meteorologie, Geophysik und Bioklimatologie, Ser. B, 18,* 253–268.

Dixey, E., 1924. Lake level in relation to rainfall and sunspots. *Nature, 114,* 659–661.

Eddy, J. A., 1983. Keynote address: An historical review of solar variability, weather and climate. In *Weather and Climate Responses to Solar Variations* (B. M. McCormac, ed.), Colorado Associated University Press, Boulder.

Frolow, V., 1933. Eleven-year cycle of the Nile and of the sun. *Comptes Rendus, 197,* 852–854.

Gage, K. S. and G. C. Reid, 1981. Solar variability and the secular variation in the tropical tropopause. *Geophysical Research Letters, 8,* 187–190.

Gerety, E. J., J. M. Wallace, and C. S. Zereos, 1977. Sunspots, geomagnetic indices and weather: A cross-spectral analysis between sunspots, geomagnetic activity, and global weather data. *Journal of the Atmospheric Sciences, 34,* 673.

Hameed, S., 1984. Fourier analysis of Nile flood levels. *Geophysical Research Letters, 1,* 843–845.

Hameed, S. and P. Wyant, 1982. Twenty-three-year cycle in surface temperatures during the Maunder Minimum. *Geophysical Research Letters, 9,* 83–86.

Hanzlik, S., 1937. The Hale double solar-cycle rainfall in western Canada. *Bulletin of the American Meteorological Society, 18,* 60–63.

Hassan, F. A., 1981. Historical Nile floods and their implications for climatic change. *Science, 212,* 1142–1145.

Hutchins, D.E., 1889. *Cycles of Drought and Good Seasons in South Africa.* Times Office, Wynberg, 136 pp.

Krick, I. P. et al., 1954. Is the drought over? *Farm Journal, Philadelphia, 78,* 31, 67.

Kullmer, C. J., 1943. A remarkable reversal in the distribution of storm frequency in the United States in the double Hale solar cycle, of interest in long-range forecasting. *Smithsonian Miscellaneous Collections, 103,* No. 10, 20 pp.

Leeper, G. W., 1963. Adelaide rainfall and the double sunspot cycle. *Australian Journal of Science, 25,* 470–471.

Lush, A., 1927. Sunspots and river flow. *Proceedings of the New Zealand Society of Civil Engineering, Wellington, 14,* 184–200.

Marshall, J. R., 1972. *Precipitation patterns of the United States and sunspots.* Ph.D. Thesis, University of Kansas, Lawrence.

Miles, M. K., 1974. Variation of annual mean surface pressure over the northern hemisphere during the double sunspot cycle. *Meteorological Magazine, 103,* 93–99.

Mitchell, J. M., Jr., C. W. Stockton, and D. M. Meko, 1979. Evidence of a 22-year rhythm of drought in the western United States related to the Hale solar cycle since the 17th century. In *Solar Terrestrial Influence on Weather and Climate,* B. M. McCormac and T. A. Seliga (eds). D. Reidel, Dordrecht, pp. 125–143.

Nassau, J. J. and W. Koski, 1933. Lake Erie levels and sunspots. *Popular Astronomy, 41,* 198–202.

Perry, C. A., 1994. Solar-irradiance variations and regional precipitation fluctuations in the western USA. *International Journal of Climatology, 14,* 969–984.

Ravenstein, E. G., 1901. The lake level of Victoria Nyanza. *Geographical Journal, 18,* 403–406.

Riehl, H., M. El-Bakry, and J. Meiten, 1979. Nile river discharge. *Monthly Weather Review, 107,* 1546–1553.

Ryder, F. J., 1947. Sunspots and the levels of lakes in Minnesota. *Popular Astronomy, 55,* 204–209.

Shukow, M. S., 1929. Discharge amount from the rivers of Central Asia and the solar activity [in German]. *Meteorological Institut Arbeiten Hydrometeorologische Abteilung, Tashkent, 1,* Part 2, 49–78.

Stewart, B., 1882. On a supposed connection between the heights of the rivers and the number of sunspots on the sun. *Nature, 26,* 489.

Streiff, A., 1929. The practical importance of climatic cycles in engineering. *Monthly Weather Review, 57,* 405–411.

Sun, C. and B. Yang, 1992. The solar activity and the regularities of droughts and waterloggings in the middle and lower reaches of the Yangtze river. *Acta Astronomica Sinica, 33,* 179–185.

Vincent, C. E., T. D. Davies, and A. K. C. Beresford, 1979. Recent changes in the level of Lake Naivasha, Kenya, as an indicator of equatorial westerlies over East Africa. *Climatic Change, 2,* 175–189.

Walker, G. T., 1936. The variations of level in lakes: Their relations with each other and with sunspot numbers. *Quarterly Journal of the Royal Meteorological Society, 62,* 451–454.

Wilson, B. H., 1946. Some deductions made from a correlation of Lake Michigan water levels with the sunspot cycle. *Popular Astronomy, 54,* 73–84.

Wood, C. A. and R. R. Lovett, 1974. Rainfall, drought and the solar cycle. *Nature, 251,* 594–596.

Chapter 7. Storms

Blanford, H. F., 1891. The paradox of the sunspot cycle. *Nature, 43,* 583–587

Brooks, C. E. P., 1934. The variation of annual frequency of thunderstorms in relation to sunspots. *Quarterly Journal of the Royal Meteorological Society, 60,* 153–162.

Brown, G. M. and J. I. John, 1979. Solar cycle influences in tropospheric circulation. *Journal of Atmospheric and Terrestrial Physics, 41,* 43–52.

Clough, H. W., 1933. The 11-year sun-spot period, secular periods of solar activity, and synchronous variations of terrestrial phenomena. *Monthly Weather Review, 60,* 99–108.

Cohen, T. J. and E. I. Sweetser, 1975. The 'spectra' of the solar cycle and of data for Atlantic tropical cyclones. *Nature, 256,* 295–296.

Herman, J. R. and R. A. Goldberg, 1978. Initiation of non-tropical thunderstorms by solar activity. *Journal of Atmospheric and Terrestrial Physics, 40,* 121–134.

Herman, J. R. and R. A. Goldberg, 1978. *Sun, Weather, and Climate.* NASA SP-426, GPO, Washington, D.C., 360 pp.; and 1985, Dover Publications, NY.

Lethbridge, M. D., 1979. Thunderstorm frequency and solar sector boundaries. In *Solar-Terrestrial Influences on Weather and Climate* (B. M. McCormac and T. A. Seliga, eds.), D. Reidel, Dordrecht, 253–258.

Lockyer, J. N., 1904. Simultaneous solar and terrestrial changes. *Nature, 69,* 351.

Markson, R., 1971. Considerations regarding solar and lunar modulation of geophysical parameters, atmospheric electricity, and thunderstorms. *Pure and Applied Geophysics, 84,* 161–202.

Meldrum, C., 1872. On a periodicity in the frequency of cyclones in the Indian Ocean south of the equator. *Nature, 6,* 357–358.

Meldrum, C., 1872. On a periodicity in the frequency of cyclones in the Indian Ocean south of the equator. *British Association Report, 42,* 56–58.

Meldrum, C., 1878. Sunspots and rainfall. *Nature, 18,* 564–567.

Meldrum, C., 1885. On a supposed periodicity of the cyclones of the Indian Ocean south of the equator. *British Association Report,* 925–926.

Merrill, R. T., 1987. *An Experiment in Statistical Prediction of Tropical Cyclone Intensity Change.* Miami, NOAA Technical Memo. NWS NHC 34.

Mukherjee, A. K. and B. P. Singh, 1978. Trends and periodicities in annual rainfall in monsoon areas over the northern hemisphere. *Indian Journal of Meteorology, Hydrology, and Geophysics, 29,* 1 and 2, 441–447.

Pittock, A. B., 1978. A critical look at long-term sun-weather relationships. *Reviews of Geophysics and Space Physics, 16,* 400–420.

Poey, M. A., 1873. Sur les rapports entre les taches solaires et les ourages des Antilles de l'Atlantique-nord et de l'Ocean Indien sud. *Comptes Rendus, 77,* 1223–1226.

Poey, M. A., 1873. Sur les rapports entre les taches solaires, les ourages a Paris et a Fecamp, les temptes et les coups de vent dans l'Atlantique-nord. *Comptes Rendus 77,* 1343–1346.

Roederer, J. G. , in press. *Paleoclimate Research, 18.*

Stringfellow, M. F., 1974. Lightning incidence in Britain and the solar cycle. *Nature, 249,* 332–333.

Visher, S. S., 1924. Sunspots and the frequency of tropical cyclones. *American Journal of Science, 8,* 312–316.

Chapter 8. Biota

Auclair, A. N. D., 1992. Forest wildfire, atmospheric CO_2, and solar irradiance periodicity. *Eos, 73,* 70.

Coope, G. R., 1977. Fossil coleopteran assemblages as sensitive indicators of climatic changes during the Devensian (last) cold stage. *Philosophical Transactions of the Royal Society, London, 280,* 313–340.

Criddle, N., 1932. The correlation of the sunspot periodicity with grasshopper fluctuations in Manitoba. *Canadian Field Nature, 46,* 195.

Currie, R. G., 1988. Lunar tides and the wealth of nations. *New Scientist,* November 5, 52–55.

Currie, R. G., 1991. Deterministic signals in tree-rings from the Corn Belt region. *Annales de Geophysicae, 9,* 565–570.

Currie, R. G., T. Wyatt, and D. P. O'Brien, 1993. Deterministic signals in European fish catches, wine harvests, and sea-level, and further experiments. *International Journal of Climatology, 13,* 665–687.

DeLury, R. E., 1923. Arrival of birds in relation to sunspots. *The Auk, 40,* 414–419.

Druzhinin, I. P. and N. V. Khamyoava, 1973. *Solar Activity and Sudden Changes in the Natural Processes on Earth: A Statistical Analysis.* NASA TT F-652, 295 pp.

Foukal, P. and J. Lean, 1990. An empirical model of total solar irradiance variation between 1874 and 1988. *Science, 247,* 556–558.

Garcia-Mata, C. and F. I. Shaffner, 1934. Solar and economic relationships, a preliminary report. *Quarterly Journal of Economics, 49,* 1–51.

Gloyne, R. W., 1973. The "growing season" at Eskdalemuir observatory, Dumfriesshire. *Meteorological Magazine, 102,* 174.

Hahn, F. G., 1877. *Über die Beziehungen der Sonnenfleckenperiode zu meteorologischen Erscheinungen.* Leipzig.

Harris, J. A., 1926. The correlation between sunspot numbers and tree growth. *Monthly Weather Review, 54,* 13–14.

Headlee, T. J., 1943. In *Solar Relations to Weather and Climate* by H. H. Clayton, Clayton Weather Service, Canton, Mass., 77–79.

Huntington, E. and S. S. Visher, 1922. *Climatic Changes.* Yale University Press, New Haven.

Izhevshii, G. K., 1964. *Systematic Principles of Forecasting Oceanological Conditions and Reproduction of Commercial Fishes* [in Russian]. All-Union Scientific Research Institute for Marine Fishing Industry and Oceanography, Moscow.

Jevons, W. J., 1875. The influence of the sunspot period upon the price of corn. *Nature, 13,* 15.

Jevons, W. J., 1878. Commercial crises and sunspots. *Nature, 19,* 588–590.

Jevons, W. J., 1878. Sunspots and commercial crises. *Nature, 19,* 33–37.

Johnson, L. P. V., 1940. Relation of sunspot periodicity to precipitation, temperature, and crop yields in Alberta and Saskatchewan. *Canadian Journal of Research, 18,* 79–91.

Keith, L. B., 1963. *Wildlife's Ten-Year Cycle.* The University of Wisconsin Press, Madison, 201 pp.

LaMarche, V., Jr. and H. C. Fritts, 1972. Tree-rings and sunspot numbers. *Tree Ring Bulletin, 32,* 19–33.

Legrand, J. P. and P. Simon, 1979. Exceptional meteorological fluctuations in Europe since 1393 selected by dates of the wine harvest and their relationship to solar activity. In *Evolution of Planetary Atmospheres and Climatology of the Earth.* CNES, Toulouse, 325–339.

Ljungman, A., 1880. Bidrag till losningen af fragen om de stora sillfiskenas sekulara periodicitet. Copenhagen. Translated by H. Jacobson in the *Report of the Commissioner for 1879,* U. S. Commission of Fish and Fisheries, Part 7, 497–503.

MacLagen, D. S., 1940. Sunspots and insect outbreaks: An epidemic-logical study. *Proceedings of the University of Durham Philosophical Society, 10,* 173–190.

MacLulich, D.A., 1936. Sunspots and abundance of animals. *Journal of the Canadian Royal Astronomical Society, 30,* 233–246.

Poland, H., 1892. *Fur-bearing Animals.* Guerney and Jackson, London, 392 pp.

Rankin, W., 1995. Astrology for snowshoe hares. *National Wildlife, 33,* 16.

Swinton, A. H., 1882. Table of the appearance of rare lepidoptera in this country in connection with the sunspot cycle. *Nature, 25,* 584.

Walker, F. T., 1956. Periodicity of the Laminariaceae around Scotland. *Nature, 177,* 1246.

Wyatt, T., R. G. Currie, and F. Saborido-Rey, 1994. Deterministic signals in Norwegian cod records. *ICES Marine Science Symposium, 198,* 49–55.

Wylie, C. C., 1927. Solar cycle in temperature and in crops. *Popular Astronomy, 35,* 253–256.

Zimmer, C., 1993. Pacemaker of the hares. *Discover, 14,* no. 6, 20.

Chapter 9. Cyclomania

Clayton, H. H., 1894. Six and seven day weather periods. *American Journal of Science, 3rd Ser., 47,* 223–231.

Lockyer, J. N., 1874. *Contributions to Solar Physics.* Macmillan, London.

Marvin, C. F., 1930. Are meteorological sequences fortuitous? *Monthly Weather Review, 58,* 490–493.

Pittock, A. B., 1978. A critical look at long-term sun-weather relationships. *Reviews of Geophysics and Space Physics, 16,* 400–420.

Pittock, A. B., 1983. Solar variability, weather, and climate: An update. *Quarterly Journal of the Royal Meteorological Society, 109,* 23–55.

Chapter 10. Solar and Climate Changes on Long Time Scales

Abarbanell, C. and H. Wohl, 1981. Solar rotation velocity as determined from sunspot drawings of J. Hevelius in the 17th century. *Solar Physics, 70,* 197–203.

Angell, J. K. and J. Korshover, 1977. Estimate of global change in temperature, surface to 100 mb, between 1958 and 1975. *Monthly Weather Review, 105,* 375–385.

Baliunas, S. And W. Soon, 1995. Are variations in the length of the activity cycle related to changes in brightness in solar-type stars? *Astrophysical Journal, 450,* 896–901.

Beer, J., F. Joos, Ch. Lukasczyk, W. Mende, J. Rodriguez, U. Siegenthaler, and R. Stellmacher, 1993. ^{10}Be as an indicator of solar variability and climate. In *The Solar Engine and Its Influence on Terrestrial Atmosphere and Climate.* Springer-Verlag, Berlin, 221–234.

Beer, J., U. Siegenthaler, G. Bonani, R. C. Finkel, H. Oeschager, M. Suter, and W. Wolfli, 1988. Information on past solar activity and geomagnetism from ^{10}Be in the Camp Century ice core. *Nature, 331,* 675–679.

Bradley, R. S. (ed.), 1992. *Climate Since* A.D. *1500.* Routledge, New York 679 pp.

Bradley, R. S. and P. D. Jones, 1993. 'Little Ice Age' summer temperature variations: Their nature and relevance to recent global warming trends. *The Holocene, 3,* 367–376.

Bray, J. R., 1965. Forest growth and glacier chronology in north-west North America in relation to solar activity. *Nature, 205,* 440–443.

Bruckner, E., 1890. Klimaschwankungen seit 1700 [Climate variations since 1700]. *Geographische Abhandlungen, 14,* 325 pp.

Budyko, M. I., 1969. The effect of solar radiation variations on the climate of the Earth. *Tellus, 21,* 611–619.

Clough, H. W., 1905. Synchronous variations in solar and terrestrial phenomena. *Astrophysical Journal, 22,* 42–75.

Clough, H. W., 1933. The 11-year sun-spot period, secular periods of solar activity, and synchronous variations of terrestrial phenomena. *Monthly Weather Review, 60,* 99–108.

Clough, H. W., 1943. The long-period variations in the length of the 11-year solar period, and on current variations in terrestrial phenomena. *Bulletin of the American Meteorological Society, 24,* 154–163.

Eddy, J. A., P. A. Gilman, and D. E. Trotter, 1974. Solar rotation during the Maunder Minimum. *Solar Physics, 46,* 3–14.

Eddy, J. A., 1976. The Maunder minimum. *Science, 192,* 1189–1192.

Eddy, J. A. and A. A. Boornazian, 1979. Secular decrease in the solar diameter, 1863–1953. *Bulletin of the American Astronomical Society, 11,* 437.

Endal, A. S., S. Sofia, and L. W. Twigg, 1985. Changes of solar luminosity and radius following secular perturbations in the convective envelope. *Astrophysical Journal, 290,* 748–757.

Friis-Christensen, E. and K. Lassen, 1991. Length of the solar cycle: An indicator of solar activity closely associated with climate. *Science, 254,* 698–700.

Gates, W. L. and Y. Mintz, 1975. *Understanding Climate Change,* Appendix A, National Academy of Sciences, Washington, D. C.

Gleissberg, W., 1966. Ascent and descent in the eighty-year cycles of solar activity. *Journal of the British Astronomical Society, 76,* 265–270.

Goad, J., 1686. *Astrometeorologica, or Aphorisms and Discourses on the Bodies Celestial, Their Nature and Influences.* Printed by J. Rawlings for O. Blagrave, London, 509 pp.

Groveman, B. S. and H. E. Landsberg, 1979. Simulated northern hemisphere temperature departures 1579–1880. *Geophysical Research Letters, 6,* 767–769.

Hansen, J. E. and S. Lebedeff, 1988. Global surface air temperatures: Update through 1987. *Geophysical Research Letters, 15,* 323.

Hoyt, D. V., 1979. Variations in sunspot structure and climate. *Climatic Change, 2,* 79–92.

Hoyt, D. V., 1990. Using the boundary conditions of sunspots as a technique for monitoring solar luminosity. In *Climate Impact of Solar Variability* (K. H. Schatten and A. Arking, eds.) NASA Conference Publication 3086, 42–49.

Hoyt, D. V. and K. H. Schatten, 1993. A discussion of plausible solar irradiance variations, 1700–1992. *Journal of Geophysical Research, 98A,* 18895–18906.

Huntington, E. and S. S. Visher, 1922. *Climatic Changes.* Yale University Press, New Haven.

Kanda, S., 1933. Ancient records of sunspots and auroras in the Far East and the variation of the period of solar activity. *Proceedings of the Imperial Academy, Japan, 9,* 293–296.

Landsberg, H. E. and R. E. Kaylor, 1976. Spectral analysis of long meteorological series. *Journal of Interdisciplinary Cycle Research, 7,* 237.

Landsberg, H. E., 1980. Variable solar emissions, the 'Maunder Minimum', and climate Temperature Fluctuations. *Archive fur Meterologie, Geophysik, und Bioklimatolgie,* Ser. B, *28,* 181–191.

Lean, J., A. Skumanich, and O. White, 1992. Estimating the Sun's radiative output during the Maunder Minimum. *Geophysical Research Letters, 19,* 1591–1594.

Lockyer, N. and W. J. S. Lockyer, 1901. On solar changes of temperature and variations of rainfall in the region surrounding the Indian Ocean. I. *Nature, 63,* 107–109.

Lockyer, N. and W. J. S. Lockyer, 1901. On solar changes of temperature and variations of rainfall in the region surrounding the Indian Ocean. II. *Nature, 63,* 128–133.

Luby, W. A., 1942, The mean period of sunspots. *Popular Astronomy, 50,* 537.

Moreno-Insertis, F. and M. Vasquez, 1988. A statistical study of the decay phase of sunspot groups from 1874 to 1939. *Astronomy and Astrophysics, 205,* 289–296.

Muller, J., 1926. Der Parallelismus zwischen der gesetzmässigen Wiederkehr strenger Winter und den Sonnenfleckenperioden von kurzer oder langerer Dauer. *Petermanns Mittheilungen 1926,* 241–247.

Nesmes-Ribes, E. and A. Manganey, 1992. On a plausible physical mechanism linking the Maunder Minimum to the Little Ice Age. *Radiocarbon, 34,* 263–270.

Nordo, J., 1955. A comparison of secular changes in terrestrial climate and sunspot activity. *Videnskaps-Akademeits Institut for Vaer-ag Klimatorskning, Oslo,* Rept. No. 5, 14 pp.

Pfister, C., 1995. Personal communication. See Wanner, H., C. Pfister, R. Brazdil, P. Frich, K. Frydendahl, T. Jonsson, J. Kington, H. H. Lamb, S. Rosenorn, and E. Wishman, 1995. Wintertime European circulation patterns during the late Maunder Minimum (1675–1704). *Theoretical and Applied Climatology, 51,* 167–175.

Reid, G. C., 1991. Solar total irradiance variations and the global sea surface temperature record. *Journal of Geophysical Research, 96,* 2835–2844.

Ribes, E., 1990. Astronomical determinations of the solar variability. *Philosophical Transactions of the Royal Society, London, 330,* 487–497.

Rind, D. and J. T. Overpeck, 1993. Hypothesized causes of decade-to-century-scale climate variability—Climate model results. *Quaternary Science Reviews, 12,* 357–374.

Sakurai, K., 1977. Equatorial solar rotation and its relation to solar activity. *Nature, 269,* 401–402.

Siegenthaler, U. and H. Oeschger, 1987. Biospheric CO_2 emissions during the past 200 years reconstructed by deconvolution of ice cores. *Tellus, 39B,* 140–154.

Socher, H. V., 1939. A long cycle in sunspot numbers. *Observatory, 62,* 277–279.

Youji, D., L. Baorong, and F. Yongming, 1979. Periodic peaks of ancient solar activities. In *Weather and Climate Responses to Solar Variations* (B. M. McCormac, ed.). Colorado Associated University Press, Boulder, 545–557.

Chapter 11. Alternative Climate Change Theories

Charlson, R. J., J. Langer, H. Rodhe, C. B. Leovy, and S. G. Warren, 1991. Perturbation of the northern hemisphere radiative balance by backscattering from anthropogenic sulfate aerosols. *Tellus, 43AB,* 152–163.

Conway, T. J., P. P. Tans, L. S. Waterman, K. W. Thorning, D. R. Kitzis, K. A. Masarie, and N. Zhang, 1995. Evidence for interannual variability of the carbon cycle from the National Oceanic and Atmospheric Administration / Climate Monitoring and Diagnostics Laboratory, Global Air Sampling Network. *Journal of Geophysical Research, 99,* 22831–22855.

Dogniaux, R. and R. Sneyers, 1972. *Sur la stabilite du trouble atmospherique a Uccle au cours de la periode 1951–1970.* Pub. Ser. A, No. 75. Institute Royal Meterologique de Belgique Brussels.

Flagg, W., 1881. *A Year Among the Trees.* Educational Publishing, Boston, 308 pp.

Franklin, B., 1785. Meteorological imaginations and conjectures. *Proceedings of the Manchester Literary and Philosophical Society, 2,* 375.

Friis-Christensen, E. and K. Lassen, 1991. Length of the solar cycle: An indicator of solar activity closely associated with climate. *Science, 254,* 698–700.

Hansen, J. E. and A. A. Lacis, 1990. Sun and dust versus greenhouse gases: An assessment of their relative roles in global climate change. *Nature, 346,* 713–719.

Hansen, J., A. Lacis, R. Ruety, M. Sato, and H. Wilson, 1993. Global climate change. *National Geographic Research Explorer, 9,* 143–158.

Harington, C. R. (ed.), 1992. *The Year without a Summer? World Climate in 1816.* Canadian Museum of Science, Ottawa. 576 pp.

Hogg, H. S., 1946. Dearth of sunspots in the seventeenth century. *Royal Astronomical Society, Canada, 40,* 373.

Hoyt, D. V. and C. Frohlich, 1983. Atmospheric transmission at Davos, Switzerland, 1909–1979. *Climatic Change, 5,* 61–72.

Hoyt, D. V. and K. H. Schatten, 1993. A discussion of plausible solar irradiance variations, 1700–1993. *Journal of Geophysical Research, 98A,* 18896–18906.

Mass, C. F. and D. A. Portman, 1989. Major volcanic eruptions and climate: A critical evaluation. *Journal of Climate, 2,* 566–593.

McWilliams, S., 1973. *Atmospheric Turbidity at Valentia Observatory.* Tech. Note no. 36, Meteorological Service, Dublin, 16 pp.

Neckel, H. and D. Labs, 1984. The solar radiation between 3300 and 12500 Å. *Solar Physics, 90,* 205–258.

Ruskin, J., 1884. *The Storm Cloud of the Nineteenth Century.* Lectures delivered at the London Institution, February 4 and 11, 1884.

Schlesinger, M. E. and N. Ramankutty, 1992. Implications for global warming of intercycle solar irradiance variations. *Nature, 360,* 330–333.

White, G., 1906. *The Natural History and Antiquities of Selborne.* J. M. Dent & Sons, London, 255 pp.

Chapter 12. Gaia or Athena? The Early Faint Sun Paradox

Lovelock, J. E., 1979. *Gaia: A New Look at Life on Earth.* Oxford University Press, Oxford, 157 pp.

Ruskin, J., 1869. *The Queen of the Air.* Metropolitan Publishing, New York, 247 pp.

Schatten, K. H. and A. S. Endal, 1982. The faint young sun-climate paradox: Volcanic influences. *Geophysical Research Letters, 9,* 1309–1311.

Schatten, K. H. and A. S. Endal, 1983. The faint young sun—climate paradox: Crustal influences. In *Weather and Climate Responses to Solar Variations* (B. M. McCormac, ed.). Colorado Associated University Press, Boulder, 626 pp.

Ulrich, R. K., 1975. Solar neutrinos and variations in the solar luminosity. *Science, 190,* 619–624.

Papers and Books to 1900

Theophrastus, 374–287 B.C. *De signis aquarum et ventorum.*

Aratus of Soli, ca. 315–245 B.C. *Diosemea.*

Virgil, 70–19 B.C. *Georgics.*

Pliny the Elder, 23–79. *Historie of the World.*

Digges, L., 1555. *A Prognostication of Right Good Effect.*

Hill, T., 1568. *The Profitable Arte of Gardening.*

Flamery, I., 1591. *Perpetuall and Naturall Prognostications of the Change of Weather.*

Godfridus, 1619. *Knowledge of Things Unknowne.*

Riccioli, J. B., 1651. *Almagestrum Novum.* Bononiae, Italy, 2 vols., 763 pp + 676 pp.

Kircher, A., 1671. *Iter Exstaticum Coeleste.* Herbipoli.

Herschel, W., 1796. Some remarks on the stability of the light of the sun. *Philosophical Transactions of the Royal Society, London,* 166.

Herschel, W., 1801. Observations tending to investigate the nature of the sun, in order to find the causes or symptoms of its variable emission of light and heat. *Philosophical Transactions of the Royal Society, London,* 265.

Herschel, W., 1801. Additional observations tending to investigate the symptoms of its variable emission of light and heat of the sun, with trials to set aside darkening glasses, by transmitting the solar rays through liquids and a few remarks to remove objections that might be made against some of the arguments contained in the former paper. *Philosophical Transactions of the Royal Society, London,* 354.

Flaugergues, 1823. Zach Correspondance Astronomique, 9.

Gruithuisen, Fr.v.P., 1826. Naturwissenschaftlicher Reisebericht. *Kastners Archiv für die gesammte Naturlehre, 8,* Nuremberg.

Capocci, 1827. Üeber die Sonnenflecken. *Astronomische Nachrichten,* no. 115, 314.

Arago, 1830. Supposed influence of the spots of the sun on temperature. *EnS, 2,* 66.

Buys-Ballot, 1844. Einfluss der Rotation der Sonne auf die Temperatur unserer Atmosphare. *Poggendorf's Annalen der Physik und Chemie, 68.*

Gautier, A., 1844. Recherches relatives à l'influence que le nombre et la permanence des taches observées sur le disque du soleil peuvent exercer sur les températures terrestres. *Annales de Chemie et de Physique, 3,* 12.

Nervander, 1844. Bull. de la classe phys. math. de l'Acadamie de Petersburg, III, *Poggendorf's Annalen der Physik und Chemie, 68.*

Schwabe, A.N., 1844. Sonnen-Beobachtungen im Jahre 1843. *Astronomische Nachrichten, 21*, 233.

Sabine, E., 1852. On periodical laws discoverable in the mean effects of the larger magnetic disturbances. *Philosophical Transactions of the Royal Society, London,* 103–124.

Fritsch, K., 1854. *Über das Steigen und Fallen der Lufttemperatur binnen einer analogen elfjährigen Periode, in welcher die Sonnenflecke sich vermindern oder vermehren,* Vienna.

Smyth, C. P., 1856. Note on the constancy of solar radiation. *Royal Astronomical Society, Memoirs, 25,* 125–131.

Wolf, R., 1859. Über den Einfluss der Sonnenflecken auf die Temperatur. *Mitteilungen über die Sonnenflecken,* no. 9, 211–214.

Greg, R. P., 1860. On the periodicity of the solar spots, and induced meteorological disturbances. *Philosophical Magazine, 20,* 4th ser., 246–247.

Buchan, A., 1867. On the cold weather of March 1867, as illustrating the relation between temperature and pressure of the atmosphere. *Journal of the Scottish Meteorological. Society, New Series, 2,* 66–78.

Stewart, B., 1869. On the sun as a variable star. *Royal Institution, Proceedings, 5,* 138.

Smyth, C. P., 1870. On supra-annual cycles of temperature in the earth's surface-crust. *Royal Society Proceedings, 18,* 311–312. Also published in *Philosophical Magazine, 40,* 58–59.

Anonymous, 1871. On the connection between terrestrial temperature and sun-spot phenomena. *Nature, 5,* 434.

Baxendall, J., 1871. On solar radiation. *Memoirs of the Manchester Literary and Philosophical Society, 3rd Ser., 4,* 128–134; 147–155.

Smyth, C. P., 1871. Solar science and the pleasure of secret referees. *Nature, 5,* 468.

Stone, E. J., 1871. On an approximately decennial variation of the temperature at the Observatory at the Cape of Good Hope between the years 1841 and 1870, viewed in connexion with the variation of the solar spots. *Royal Society Proceedings, 19,* 389–392. Also published in *Philosophical Magazine, 42,* 72–75.

Anonymous, 1872. Cholera and sun-spots. *Nature, 7,* 26.

Meldrum, C., 1872. On a periodicity in the frequency of cyclones in the Indian Ocean south of the equator. *Nature, 6,* 357–358. Also published in *British Association Report, 42,* 56–58.

Schuster, A., 1872. Sun-spots and the vine crop. *Nature, 6,* 501.

Koppen, W., 1873. On temperature cycles. *Nature, 9,* 184–185.

Koppen, W., 1873. Über vieljährige Perioden der Witterung. *Deutsche Rundschau für Geographie und Statistik, 2.*

Koppen, W., 1873. Über mehrjährige Perioden der Witterung, insbesondere über die 11-jährige Periode der Temperatur. *Zeitschrift der Österreichischen Gesellschaft für Meteorologie, 8,* 241–248, 256–267.

Meldrum, C., 1873. On a periodicity of cyclones and rainfall in connexion with the sunspot periodicity. *British Association Report, 43,* 466–478. Also published in *Nature, 8,* 495.

Meldrum, C., 1873. On the periodicity of rainfall in connexion with the sun-spot periodicity. *Royal Society Proceedings, 21,* 297–308.

Meldrum, C., 1873. On a supposed periodicity of the rainfall in Mauritius and Australia. *Quarterly Journal of the Royal Meteorological Society, 1,* 130–133.

Meldrum, C., 1873. Periodicity of rainfall. *Nature, 8,* 547–548.

Poey, M. A., 1873. Sur les rapports entre les taches solaires et les ourages des Antilles de l'Atlantique-nord et de l'Ocean Indien sud. *Comptes Rendus, 77,* 1223–1226.

Poey, M. A., 1873. Sur les rapports entre les taches solaires, les ourages a Paris et a Fecamp, les temptes et les coups de vent dans l'Atlantique-nord. *Comptes Rendus, 77,* 1343–1346.

Rawson, R. W., 1873. Periodicity of rainfall. *Nature, 8,* 245.

Dawson, G. M., 1874. The fluctuations of the American lakes and the development of sun-spots. *Nature, 9,* 504.

Meldrum, C., 1874. On cyclone and rainfall periodicities in connexion with the sunspot periodicity. *British Association Report,* 218–240. Also published in *Nature, 10,* 431.

Meldrum, C., 1874. Periodicity of rainfall. *Nature, 11,* 327.

Moffat, T., 1874. On the apparent connection between sun-spots and atmospheric ozone. *Nature, 10,* 411.

Blanford, H. F., 1875. On some recent evidence of the variation of the Sun's heat. *Journal of the Asiatic Society of Bengal, 44,* 21–35, 120–122.

Blanford, H. F., 1875. Solar heat and sun-spots. *Nature, 11,* 147–148.

Blanford, H. F., 1875. Solar radiation and sun-spots. *Nature, 11,* 188–189. Also published in *Meteorologische Zeitung, 10,* 261–264.

Chambers, C., 1875. *Meteorology of the Bombay Presidency,* Bombay.

Jevons, W. S., 1875. The influence of the sunspot period upon the price of corn. *Nature, 13,* 15.

Moffat, T., 1875. On the apparent connection between sunspots, atmospheric ozone, and the force of the wind. *Nature, 12,* 374.

Secchi, P. A., 1875. Ericsson's researches on the sun. *Nature, 13,* 46.

Langley, S. P., 1876. Measurement of the direct effect of sun-spots on terrestrial climates. *Monthly Notices of the Royal Astronomical Society, 37,* 5–11.

Meldrum, C., 1876. On cyclone and rainfall periodicities in connexion with the sunspot periodicity. *British Association Report,* 267–274.

Meldrum, C., 1876. On a secular variation in the rainfall in connexion with the secular variation in amount of sunspots. *Royal Society Proceedings, 24,* 379–387.

Meldrum, C., 1876. Sunspots and the prediction of the weather of the coming season at Mauritius. *Monthly Notices of the Metetereological Society of Mauritius, 1,* 1.

Williams, W. M., 1876. Note on Prof. Langley's paper on the direct effect of sunspots on terrestrial climates. *Monthly Notices of the Royal Astronomical Society, 37,* 41–42.

Anonymous, 1877. Sunspots and famines. *English Mechanic, 26,* 380.

Anonymous, 1877. Sunspots and rainfall. *Nature, 17,* 443–445.

Archibald, E. D., 1877. Sunspots and rainfall at Calcutta. *Nature, 16,* 267.

Archibald, E. D., 1877. Relations between sun and earth. *Nature, 16,* 339–341.

Archibald, E. D., 1877. Indian rainfall and sunspots. *Nature, 16,* 396–397.

Archibald, E. D., 1877. Rainfall and sunspots in India. *Nature, 16,* 438–439.

Archibald, E. D., 1877. Indian rainfall. *Nature, 17,* 505.

Bonovia, E., 1877. Sunspots and rainfall. *Nature, 17,* 443–445.

Boreas, R., 1877. Sunspots and weather. *English Mechanic, 25,* 408.

Broun, J. W., 1877. Rainfall and sunspots. *Nature, 16,* 251–252.

Chambers, C., 1877. The meteorology of the Bombay presidency. *Royal Society Proceedings, 25,* 539–540.

Fritz, H., 1877. Sunspots and weather. *Nature, 15,* 263.

Hill, S. A., 1877. Rainfall in the temperate zone in connection with the sunspot cycle. *Nature, 17,* 59–60.

Hunter, W. W., 1877. Relations between sun and earth. *Nature, 16,* 359.

Hunter, W. W., 1877. The cycle of sunspots and rainfall. *Nature, 16,* 455–456.

Jeula, H., 1877. Sunspots and ship wrecks. *Nature, 16,* 447.

Meldrum, C., 1877. Sunspots and rainfall. *Nature, 17,* 448–450. Also published in *British Association Report,* 230–258.

Stewart, B., 1877. Indian rainfall and sunspots. *Nature, 16,* 161.

Stewart, B., 1877. Suspected relations between the sun and the earth. I. *Nature, 16,* 9–11.

Stewart, B., 1877. Suspected relations between the sun and the earth. II. *Nature, 16,* 26–28.

Stewart, B., 1877. Suspected relations between the sun and the earth. III. *Nature, 16,* 45–47.

Archibald, E. D., 1878. Locusts and sunspots. *Nature, 19,* 145–146.

Broun, J. A., 1878. Sunspots, atmospheric pressure and the sun's heat. *Nature, 19,* 6–9.

Capello, J., 1878. Atmospheric pressure. *Nature, 19,* 506.

Chambers, F., 1878. Sun and earth. *Nature, 18,* 619–620.

Chambers, F., 1878. Sun-spots and weather. *Nature, 18,* 567–568.

Derby, D.A., 1878. The rainfall of Brazil and sunspots. *Nature, 18,* 384–385.

Fritz, H., 1878. *Die Beziehungen der Sonnenflecke zu den magnetischen und meteorologischen Erscheinungen der Erde.* Haarlem, The Netherlands.

Hill, S.A., 1878. Atmospheric pressure and solar heat. *Nature, 19,* 432.

Jevons, W. J., 1878. Commercial crises and sunspots. *Nature, 19,* 588–590.

Jevons, W. J., 1878. Sunspots and commercial crises. *Nature, 19,* 33–37.

Jevons, W. J., 1878. Sunspots and the plague. *Nature, 19,* 338.

Kemp, J., 1878. Commercial crises and sunspots. *Nature, 19,* 97.

Meldrum, C., 1878. Sunspots and rainfall. *Nature, 18,* 564–567.

Stewart, B., 1878. Sunspots and weather. *Nature, 18,* 616.

Wilson, W. S., 1878. Commercial crises and sunspots. *Nature, 19,* 196–197.

Archibald, E. D., 1879. Barometric pressure and sunspots. *Nature, 20.*

Archibald, E. D., 1879. The weather and the sun. *Nature, 20,* 626–627.

Archibald, E. D., 1879. On the connection between solar phenomena and climatic cycles. *Ramsay's Scientific Roll, 1.*

Blanford, H. F., 1879. *Report on the Meteorology of India in 1878,* Bombay.

Hill, S. A., 1879. Úber eine zehnjährige Periode in der jährlichen Änderung der Temperatur und des Luftdruckes in Nord-Indien. *Zeitschrift der Österreichischen Gesellschaft für Meteorologie, 14.*

Hirsch, 1879. Sur l'influence des taches solaires sur la temperature de la terre. *Neuchatel, Bulletin, 11,* 142.

Hosie, A., 1879. The first observations of sunspots. *Nature, 20,* 131–132.

Meldrum, C., 1879. Sunspots and the rainfall of Paris. *Nature, 21,* 166–168.

Parfitt, E., 1879. Sunspot, &c. *Nature, 21,* 324.

Blanford, H. F., 1880. On the barometric see-saw between Russia and India in the sunspot cycle. *Nature, 21,* 477–482. Also published in *Meteorologische Zeitung, 15,* 153–158.

Chambers, F., 1880. Abnormal variations of barometric pressure in the tropics, and their relation to sunspots, rainfall, and famines. I. *Nature, 23,* 88–91.

Chambers, F., 1880. Abnormal variations of barometric pressure in the tropics, and their relation to sunspots, rainfall, and famines. II. *Nature, 23,* 107–114.

Chambers, F., 1880. Abnormal variations of barometric pressure in the tropics, and their relation to sunspots, rainfall, and famines. III. *Nature, 23,* 399–400.

Hill, S.A., 1880. Variations of rainfall in Northern India, *Indian Meteorological Memoirs, I* no. 7. See also *Meteorologische Zeitung*, 1880.

Koppen, W., 1880. Über mehrjährige Perioden der Witterung, insbesondere über die 11-jährige Periode der Temperatur. *Zeitschrift der Österreichischen Gesellschraft für Meteorologie, 16,* 140–150; 183–194; 279–283.

Liznar, 1880. Beziehung der täglichen und jährlichen Temperaturschwankung zur 11-jährigen Sonnenfleckenperiode. *Sitzungsberichte der Wiener, Akademie, 82* Also published in *Nature, 23,* 1331, 1880.

Archibald, E. D., 1881. Abnormal variations of barometric pressure, pressure in the tropics, and their relations to sunspots, rainfall, and famines. *Nature, 23,* 399–400.

Archibald, E. D., 1881. Abnormal barometric gradient between London and St. Petersburg, in the sunspot cycle. *Nature, 23,* 618–619.

Archibald, E. D., 1882. Variations in the sun's heat. *Nature, 25,* 316.

Blanford, H. F., 1882. Some further results of sun-thermometer observations with reference to atmospheric absorption and the supposed variation of solar heat. *Journal of the Asiatic Society of Bengal, 51,* 72–84.

Dorbreck, W., 1882. Sunspots and Markee rainfall. *Nature, 26,* 366–367.

Stewart, B., 1882. On a supposed connection between the heights of the rivers and the number of sunspots on the sun. *Nature, 26,* 489.

Swinton, A. H., 1882. Table of the appearance of rare lepidoptera in this country in connection with the sunspot cycle. *Nature, 25,* 584.

Williams, C. J. B., 1882. Unprecedented cold in the Riviera—absence of sunspots. *Nature, 27,* 551–552.

Chambers, C., 1883. Sunspots and terrestrial phenomena. On the variations of the daily range of I. atmospheric temperature, and II. the magnetic declination, as recorded at the Cordoba Observatory, Bombay. *Royal Society Proceedings, 34,* 231–264.

Swinton, A. H., 1883. Sun-Spottery. *Journal of Science, 20,* 77.

Williams, C. J. B., 1883. On the cold in March, and absence of sunspots. *Nature, 28,* 102–103.

Woeikof, A., 1883. The weather and sunspots. *Nature, 28,* 53–54.

Clayton, H. H., 1884. A lately discovered meteorological cycle of 25-months. I. *American Meteorological Journal, 1.*

Langley, S. P., 1884. *Researches on the Solar Heat and its Absorption by the Earth's Atmosphere. A Report of the Mount Whitney Expedition.* Professional Paper 15, Signal Service, Washington, D.C.

Clayton, H. H., 1885. A lately discovered meteorological cycle of 25-months. II. *American Meteorological Journal, 1.*

Clayton, H. H., 1885. A temperature oscillation of about 30 days. *American Meteorological Journal, 2,* 87.

Meldrum, C., 1885. On a supposed periodicity of the cyclones of the Indian Ocean south of the equator. *British Association Report,* 925–926.

Chambers, F., 1886. Sunspots and prices of Indian food grains. *Nature, 34,* 100–104.

Bezold, W. V., 1888. Über eine nahezu 26tätige Periodicität der Gewittererscheinungen. *Press. Akademie der Wissenschaft, Sitzungsber. Phys. Mat. Klasse,* 905–914.

Blanford, H. F., 1889. *A Practical Guide to the Climates and Weather of India, Ceylon, and Burma and the Storms of Indian Seas.* Macmillan and Co. London, 369 pp.

Buchan, A., 1889. Report on atmospheric circulation based on the observations made on board H. M. S. Challenger during the years 1873–1876 and other meteorological observations. *Challenger Report, Physics and Chemistry, 2.*

Elliot, J., 1889. Sunspots and weather in India. *Meteorological Report for India.*

Hutchins, D.E., 1889. *Cycles of Drought and Good Seasons in South Africa.* Times Office, Wynberg, 136 pp.

Spoerer, G., 1889. Über die Periodicität der Sonnenflecken seit dem Jahre 1618. *Nova Acta der Kraiserliche Leopold-Caroline Deutschen Akademie der Naturforscher, 53,* 283–324.

Archibald, E. D., 1890. Cyclical periodicity in meteorological phenomena. *American Meteorological Journal, 7,* 289–295.

Bruckner, E., 1890. Klimaschwankungen seit 1700 [Climate variations since 1700]. *Geographische Abhandlungen, 14,* 325 pp.

Veeder, M. A., 1890. The forces concerned in the development of storms. *Rochester Academy of Science, Proceedings., 1,* 57–63.

Blanford, H. F., 1891. The paradox of the sunspot cycle in meteorology. *Nature, 43,* 583–587.

Koppen, W., 1891. Besprechung von Unterwegers Arbeit "Über die kleinen Perioden der Sonnenflecken u.s.w." *Meteorologische Zeitung, 8.*

Unterweger, J., 1891. *Über die kleinen Perioden der Sonnenflecken und ihre Beziehung zu einigen periodischen Erscheinungen der Erde.* Vienna, Austria.

MacDowall, A. B., 1892. Sunspots and air temperatures. *Nature, 45,* 271–272.

MacDowall, A. B., 1892. Thunderstorms and sunspots. *Nature, 46,* 488–489.

Meldrum, C., 1892. On the sunspots, magnetic storm, cyclones and rainfall of February, 1892. *Nature, 46,* 20–21.

Fritz, H., 1893. Die perioden solarer und terrestrischer erscheinungen. *Zurich Vierteljahreschriften, 38,* 77–107.

Fritz, H., 1893. *Die wichtigsten periodischen Erscheinungen der Meteorologie und Kosmologie.* Ph.D. Thesis, University of Leipzig.

Hall, J. P., 1893. A short cycle in weather. *American Journal of Science, 3rd Ser., 45,* 227–240.

Veeder, M. A., 1893. Periodic and non-periodic fluctuations in the latitude of storm tracks. *International Congress of Meteorology, Chicago, Trans.*

Anonymous, 1894. Sunspots and solar radiation. *Nature, 49,* 274.

Anonymous, 1894. Sunspots and weather. *Nature, 50,* 113.

Bigelow, F. H., 1894. Inversions of temperatures in the 26.68 day solar magnetic period. *American Journal of Science, 3rd Ser., 48,* 435–451.

Bigelow, F. H., 1894. The polar radiation from the sun and its influence in forming the high and low atmospheric pressures of the United States. *Astronomy and Astrophysics, 13.*

Bigelow, F. H., 1894. West Indian hurricanes and solar magnetic influence. *Astronomy and Astrophysics, 13.*

Clayton, H. H., 1894. Rhythm in the weather. *The Boston Commonwealth,* November 17.

Clayton, H. H., 1894. Six and seven day weather periods. *American Journal of Science, 3rd Ser., 47,* 223–231.

Dallas, W. L., 1894. *Indian Meteorological Memoirs, 6.*

Hansky, A., 1894. *Bulletin de l'academie Imperial Society, Petersburg, 20.*

Hazen, H. A., 1894. West Indian hurricanes and solar magnetic influence. *Astronomy and Astrophysics, 13,* 105, 443.

Pettersson, O., 1894. A review of Swedish hydrographic research in the Baltic and the North Sea. *Scottish Geographical Magazine.*

Savelief, M. R., 1894. Sur l'influence qu'exercent les taches solaires sur la quantite de chaluer recue par la terre. *Comptes Rendus, 118,* 62–63.

Veeder, M. A., 1894. *The relation between solar and terrestrial phenomena.* Lyons, N.Y., 8 pp.

Clayton, H. H., 1895. Eleven year sunspot-weather period and its multiples. *Nature, 51,* 436–437.

Lockyer, J. N., 1895. Observations of sunspot spectra. I. The widening of iron lines and unknown lines in relation to the sunspot period. *Nature, 51,* 448–449.

MacDowall, A. B., 1895. Rain in August. *Nature, 52,* 519–520.

Anonymous, 1896. Sunspots, comets, and climate variations. *Nature, 55,* 42.

MacDowall, A. B., 1896. The climate of Bremen in relation to sun-spots. *Nature, 54,* 572–573.

MacDowall, A. B., 1896. Sonnenflecken und Sommertemperaturen. *Meteorologische Zeitung, 13.*

Pettersson, O., 1896. Über die Beziehungen zwischen hydrographischen und meteorologischen Phänomenen. *Meteorologische Zeitung, 13.*

Schreiber, P., 1896. Vier Abhandlungen über Periodizität des Niederschlages, theoretische Meteorologie und Gewitterregen. *Abhandlungen der Königliche söchs. meteorologisches Institutes.* vol. I. Leipzig.

Unterweger, J., 1896. Über zwei trigonometrische Reihen fur Sonnenflecken, Kometen und Klimatschwankungen. *Denkschriften der Mathematik und Natur Wissenschaft Classe der Kaiserlichen Academie der Wissenschaft,* Vienna, *44.*

Young, C. A., 1896. *The Sun.* New York: D. Appleton.

Anonymous, 1897. Sunspots and the mean yearly temperature at Turin. *Nature, 56,* 350.

Gunther, S., 1897. *Handbuch der Geophysik,* vol 1. 2nd ed.

MacDowall, A. B., 1897. Suggestions of sunspot influence on the weather of western Europe. *Quarterly Journal of the Royal Meteorological Society, 23,* 243–250.

MacDowall, A. B., 1897. Sunspots and the weather. *Nature, 57,* 16.

Rizzo, G. B., 1897. Sulla relazione per le macchie solare e la temperatura dell' ariaa Torino.

Hilderbrandson, H.H., 1897–1914. Quelques recherches sur les centres d'action de l'atmosphére. Köngli Svenska Vetenskaps-*Akademiens Förhandlingar,* Stockholm, *39,* no. 3, 1897; *32,* no. 4, 1899; *45,* No. 2, 1909; *45,* No. 11, 1910; *51,* No. 8, 1914.

Bigelow, F. H., 1898. Report on solar and terrestrial magnetism in their relation to meteorology. *Weather Bureau Bulletin,* no. 21, Washington, D.C.

Clayton, H. H., 1898. Weather harmonics. *Science, 7,* 243–245.

Flammarion, C., 1898. Le soleil et la nature. *Bulletin de la Societe Astronomique de France.*

MacDowall, A. B., 1898. Sunspots and air temperature. *Nature, 59,* 77.

MacDowall, A. B., 1898. Sunspots and weather. *Nature, 59,* 462.

Meinardus, W., 1898. Der Zusammenhang des Winterklimas in Mittel- und Nordwesteuropa mit dem Golfstrom. *Zeitschrift der Gesellschaft für Erdkunde, 33,* Berlin.

Meinardus, W., 1898. Über einige meteorologische Beziehungen zwischen dem Nordatlantischen Ozean und Europa im Winterhalbjahr. *Meteorologische Zeitung, 25 ,* 33.

Very, F. W., 1898. The variation of solar radiation. *Astrophysical Journal, 7,* 255–272.

Gunther, S., 1899. *Handbuch der Geophysik,* vol. 2, 2nd ed.

Lockyer, N. and W. J. S. Lockyer, 1899. On solar changes of temperature and variations of rainfall in the region surrounding the Indian Ocean. *Proceedings of the Royal Society, London, 67,* 409. Also published in *Nature, 63,* 107, and in French, Les changements de la temperature solaire et les variations de la pluie dans les regions qui entrourent l'ocean Indien, in *Comptes Rendes, 131,* 928.

MacDowall, A. B., 1899. Sunspots and rainfall. *Nature, 59,* 583–584.

Pettersson, O., 1899. Hvilka äro orsakerna till vegetationsperiodens tidigare eller senare inträdande under olika ar? *Konigliche Svenska Landbruksakademiens Tidsskrift,* Stockholm.

Pettersson, O., 1899. Úber den Einfluss der Eisschmelzung auf die oceanische Cirkulation. *Konigliche Svenska Vetenskaps-Akademiens Förhandlingar,* Stockholm, No. 3.

Anonymous, 1900. Sunspots and terrestrial temperature. *Knowledge, 24,* 108.

Elvins, A., 1900. Rainfall and sun-spot period. *Journal of the British Astronomical Association, 11,* 245.

Lockyer, N. and W. J. S. Lockyer, 1900. Änderungen der Sonnentemperatur und Variationen des Regenfalls in den Landerings um den Indischen Ocean. *Meteorologische Zeitung, 18,* 352–353. Also published in French, Les changements de la temperature solaire et les variations de la pluie dans les regions qui entouresnt l'ocean Indien, in *Academie Science Comptes Rendus* (Paris), *131,* 928–929.

MacDowall, A. B., 1900. Sunspots and frost. *Nature, 62,* 599.

Meinardus, W., 1900. Einige Beziehungen zwischen der Witterung und den Ernte-Erträgen in Nord-Deutschland. *Verhandlungen des VII. Internationalen Geographen-Kongresses in Berlin.*

Pettersson, O., 1900. Om drifisen i Norra Atlanten. *Ymer* (Stockholm).

Selected Papers and Books from 1900 to 1993

Lockyer, N. and W. J. S. Lockyer, 1901. On solar changes of temperature and variations of rainfall in the region surrounding the Indian Ocean. I. *Nature, 63,* 107–109.

Lockyer, N. and W. J. S. Lockyer, 1901. On solar changes of temperature and variations of rainfall in the region surrounding the Indian Ocean. II. *Nature, 63,* 128–133.

Ravenstein, E. G., 1901. The lake level of Victoria Nyanza. *Geographical Journal, 18,* 403–406.

Abbot, C. G., 1902. The relation of the sun-spot cycle to meteorology. *Monthly Weather Review, 30,* 178.

Langley, S. P., 1903. The "solar constant" and related problems. *Astrophysical Journal, 17,* 89.

Nordmann, C., 1903. Connection between sunspots and atmospheric temperature. *Nature, 68,* 162.

Langley, S. P., 1904. On the possible variation of the solar radiation and its probable effect on terrestrial temperatures. *Astrophysical Journal, 19,* 305.

Lockyer, J. N., 1904. Simultaneous solar and terrestrial changes. *Nature, 69,* 351.

Abbot, C. G., F. E. Fowle, and L. B. Aldrich, 1908 on. *Annals of the Astrophysical Observatory of the Smithsonian Institutions,* vols. 2–7. U. S. Government Printing Office, Washington, D.C.

Bigelow, F. H., 1908. The relations between the meteorological elements of the United States and the solar radiation. *American Journal of Science, 25.*

Hale, G. E., 1908. On the probable existence of magnetic fields in sunspots. *Astrophysical Journal, 28,* 315–343.

Newcomb, S., 1908. A search for fluctuations in the sun's thermal radiation through their influence on terrestrial temperature. *Transactions of the American Philosophical Society, Philadelphia, 21,* 309–387.

Humphreys, W. J., 1910. Solar disturbances and terrestrial temperatures. *Astrophysical*

Journal, 32, 97–111.

Humphreys, W. J., 1913. Volcanic dust and other factors in the production of climatic changes, and their possible relation to ice ages. *Bulletin of the Mount Weather Observatory, 6,* part I.

Huntington, E., 1914. The solar hypothesis of climatic changes. *Bulletin of Geological Society of America, 25,* 477–590.

Koppen, W., 1914. Lufttemperaturen, Sonnenflecke und Vulkanausbruche. *Meteorologische Zeitung, 31,* 305–328.

Arctowski, H., 1915. Volcanic dust veils and climatic variations. *Annals of the New York Academy of Sciences, 24,* 149–174.

Douglass, A. E., 1915. The correlation of sunspots, weather, and tree growth. *Popular Astronomy, 23,* 601.

Walker, G. T., 1915. Correlation in seasonal weather. V. Sunspots and Temperature. *Memoirs, India Meteorology,* Dept. 21, 61–90.

Huntington, E., 1916. Terrestrial temperature and solar changes. *American Geographical Society Bulletin., 27,* 184–189.

Knott, C. G., 1916. The solar radiation constant and associated meteorological problems. *Scottish Meteorological Society Journal, 17,* 74–84.

Abbot, C. G., 1917. The sun and the weather. *The Scientific Monthly,* November 1917, 400.

Huntington, E., 1918. Solar activity, cyclonic storms, and climatic changes. *Monthly Weather Review, 45,* 411–431.

Huntington, E., 1918. Solar disturbances and terrestrial weather. I. Extreme barometric gradients compared with sunspots. *Monthly Weather Review, 46,* 123–141.

Huntington, E., 1918. Solar disturbances and terrestrial weather. II. Sunspots compared with changes in the weather. *Monthly Weather Review, 46,* 168–177.

Huntington, E., 1918. Solar disturbances and terrestrial weather. III. Faculae and the solar constant compared with barometric gradients. *Monthly Weather Review, 46,* 269–277.

Clayton, H. H., 1919. Variation in solar radiation and the weather. *Smithsonian Miscellaneous Collections, 71,* 53 pp.

Henry, A. J., 1921. A review of some of the literature on the sunspot pressure relations. *Monthly Weather Review, 49,* 281–284.

Searles, P. G., 1921. Effect of sun-spots on the terrestrial temperature. *Popular Astronomy, 30,* 222–225.

Anonymous, 1923. Sunspots and air temperature in America. *Nature, 112,* 602.

Brooks, C. E. P., 1923. Variations in the levels of the central African lakes Victoria and Albert. *Geophysical Memoirs, London, 2,* 337–334.

Chree, C., 1924. Periodicities, solar and meteorological. *Quarterly Journal of the Royal Astronomical Society, 50,* 87–97.

Brunt, D., 1925. Periodicities in European weather. *Philosophical Transactions of the Royal Society, London, 225A,* 247.

Hale, G. E. and S. B. Nicholson, 1925. The law of sun-spot polarity. *Astrophysical Journal, 62,* 270.

Brooks, C. E. P., 1926. *The relations of solar and meteorological phenomena: A summary of the literature from 1914 to 1924 inclusive, etc., first report.* Commission for the Study of Solar and Terrestrial Relationships, International Council of Scientific Unions, Brussels, 66–100.

Douglass, A. E., 1926. Solar records in tree growth. *Journal of the Canadian Royal Astronomical Society, 21,* 277–280.

Brooks, C. E. P., 1928. Periodicities in the Nile floods. *Royal Meteorological Society, Memoirs, 2,* 12, 26 pp.

Douglass, A. E., 1928. *Climatic Cycles and Tree Growth.* Carnegie Institute of Washington, Publication 289; Part II, Washington, D.C.

Kullmer, C. J., 1933. The latitude drift of the storm track in the 11-year solar period. Storm frequency maps of the United States, 1883–1930. *Smithsonian Miscellaneous Collections, 89,* 2.

Abbot, C. G., 1934. *The Sun and the Welfare of Man.* New York: Smithsonian Institution Series, Inc.

Andrews, L. B., 1936. The earth, the sun, and sunspots. *Smithsonian Institution Publication 3418,* 137–177.

Brooks, C. E. P., 1936. The relations of solar and meteorological phenomena: A summary of the results from 1925 to 1934 inclusive. ICSU, Fourth Report of the Commission for the Study of Solar and Terrestrial Relationships. Florence, 147–154.

Walker, G. T., 1936. The variations of level in lakes: Their relations with each other and with sunspot numbers. *Quarterly Journal of the Royal Meteorological Society, 62,* 451–454.

Clough, H. W., 1943. The long-period variations in the length of the 11-year solar period, and on current variations in terrestrial phenomena. *Bulletin of the American Meteorological Society, 24,* 154–163.

Moseley, E. L., 1944. Recurrence of floods and droughts after intervals of about 90.4 years. *Popular Astronomy, 52,* 284–287.

Bollinger, C. J., 1945. The 22-year solar pattern of rainfall in Oklahoma and Kansas. *Bulletin of the American Meteorological Society, 26,* 376–383.

Wilson, B. H., 1946. Some deductions made from a correlation of Lake Michigan water levels with the sunspot cycles. *Popular Astronomy, 54,* 73–84.

Craig, R. A., 1951. Solar variability and meteorological anomalies. *Proceedings of the American Academy of Arts and Sciences, 79,* 280–290.

Craig, R. A. and H. C. Willett, 1951. Solar energy variations as a possible cause of anomalous weather changes. In *Compendium of Meteorology.* American Meteorological Society, Boston, 379–390.

Bell, B., 1953. Solar variation as an explanation of climatic change. In *Climatic Change,* Harvard University Press, Cambridge, 123–126.

Nordo, J., 1955. A comparison of secular changes in terrestrial climate and sunspot activity. *Videnskaps-Akademeits Institut for Vaer-ag Klimatorskning, Oslo,* Report no. 5, 14 pp.

Anonymous, 1957. Statement on solar influences on weather. *Bulletin of the American Meteorological Society, 38,* 480–481.

Nupen, W. and M. Kageorge, 1958. *Bibliography on Solar-Weather Relationships.* American Meteorological Society, Washington, D. C., 1–248.

Sterne, T. E. and N. Dieter, 1958. The constancy of the solar constant. *Smithsonian Contributions to Astrophysics, 3,* 9.

Ellison, M. A., 1959. *The Sun and Its Influence. An Introduction to the Study of Solar-terrestrial Relations.* Routledge and Kegan Paul, Ltd., London, 237 pp.

Anonymous, 1961. Solar variations, climatic change, and related geophysical problems. *Annals of the New York Academy of Sciences, 95,* 740 pp.

Leeper, G. W., 1963. Adelaide rainfall and the double sunspot cycle. *Australian Journal of Science, 25,* 470–471.

Mitchell, J. M., Jr., 1965. The solar inconstant. *Proceedings of a Seminar on Possible*

Responses of Weather Phenomena to Variable Extraterrestrial Influences. NCAR Technical Note TN-8, 155.

Willett, H. C., 1967. Solar-climatic relationships. In *The Encyclopedia of Atmospheric Sciences and Astrogeology* (R. W. Fairbridge, ed.), Reinhold Publishing, New York, 869–878.

Dehsara, M. and K. Cehak, 1970. A global survey on periodicities in annual mean temperatures and precipitation totals. *Archiv für Meteorologie, Geophysik, und Bioklimatologie, Ser. B, 18,* 269–278.

Johnsen, S. J., W. Dansgaard, H. B. Clausen, and C. C. Langway, 1970. Climatic oscillations 1200–2000 A. D. *Nature, 227,* 482.

LaMarche, V., Jr. and H. C. Fritts, 1972. Tree-rings and sunspot numbers. *Tree Ring Bulletin, 32,* 19–33.

Marshall, J. R., 1972. *Precipitation Patterns of the United States and Sunspots.* Ph.D. Thesis, University of Kansas, Lawrence.

Currie, R. G., 1974. Solar cycle signal in surface air temperature, *Journal of Geophysical Research, 79,* 5657–5660.

Meadows, A. J., 1975. A hundred years of controversy over sunspots and weather. *Nature, 256,* 95–97.

Eddy, J. A., 1976. The Maunder minimum. *Science, 192,* 1189–1192.

Bell, G. J., 1977. Changes in sign of the relationship between sunspots and pressure, rainfall and the monsoons. *Weather, 32,* 26–32.

Eddy, J. A., 1977. Climate and the changing sun. *Climatic Change, 1,* 173–190.

Mayaud, P. N., 1977. On the reliability of the Wolf number series for estimating long-term periodicities. *Journal of Geophysical Research, 82,* 1271.

Miles, M. K. and P. B. Gildersleeves, 1977. A statistical study of the likely causative factors in the climatic fluctuations of the last 100 years. *Meteorological Magazine, 106,* 314–322.

Mitchell, J. M., Jr., C. W. Stockton, and D. M. Meko, 1977. Drought cycles in the United States and their relation to sunspot cycle since 1700 A.D. *Eos, 58,* 694.

Businger, S., 1978. *Investigation of Possible Sun-Weather Relationships.* Ph.D. Thesis, Colorado University, Boulder, 82 pp.

Pittock, A. B., 1978. A critical look at long-term sun-weather relationships. *Review of Geophysics and Space Physics, 16,* 400–420.

Currie, R. G., 1979. Distribution of solar cycle signal in surface air temperature over North America. *Journal of Geophysical Research, 84,* 753–761.

Friedman, H., 1979. Reviews of space science. I. Sun and earth: A new view of climate. *Astronautics and Aeronautics, 17,* 20–25.

Hoyt, D. V., 1979a. The Smithsonian Astrophysical Observatory solar constant program. *Reviews of Geophysics and Space Physics, 17,* 427–458.

Hoyt, D. V., 1979b. Variations in sunspot structure and climate. *Climatic Change, 2,* 79–92.

Currie, R. G., 1980. Detection of the 11-year sunspot cycle signal in Earth rotation. *Geophysical Journal of the Royal Astronomical Society, 61,* 131–140.

Willson, R. C., Duncan, C. H., and Geist, J., 1980. Direct measurement of solar luminosity variation. *Science, 207,* 177–179.

Currie, R. G., 1981. Solar cycle signal in air temperature in North America: Amplitude, gradient, phase and distribution. *Journal of the Atmospheric Sciences, 38,* 808–818.

Currie, R. G., 1981. Solar cycle signal in earth rotation: Nonstationary behavior. *Science, 211,* 386–389.

Gilliland, R. L., 1982. Solar, volcanic, and CO_2 forcing of recent climatic changes. *Climatic Change, 4,* 111–131.

Livingston, W. C., 1982. Magnetic fields, convection and solar luminosity variability. *Nature, 297,* 208–209.

Livingston, W. C. and H. Holweger, 1982. Solar luminosity variation. IV. The photospheric lines, 1976–1980. *Astrophysical Journal, 252,* 375–385.

Newkirk, G., 1983. Variations of solar luminosity. *Annual Reviews of Astronomy and Astrophysics, 21,* 429–467.

Ribes, E., J. C. Ribes, and R. Bartholot, 1987. Evidence for a larger sun with a slower rotation during the seventeenth century. *Nature, 326,* 52–55.

Schatten, K. H., 1988. A model for solar constant secular changes. *Geophysical Research Letters, 15,* 121–124.

Willson, R. C. and H. S. Hudson, 1988. Solar luminosity variations in solar cycle 21. *Nature, 332,* 810–812.

Lean, J., 1989. Contribution of ultraviolet irradiance variations to changes in the sun's total irradiance. *Science, 244,* 197–200.

Baliunas, S. and R. Jastrow, 1990. Evidence for long-term brightness changes of solar-type stars. *Nature, 348,* 520–523.

Foukal, P., 1990. The variable sun. *Scientific American, 262,* 34–41.

Foukal, P. and J. Lean, 1990. An empirical model of total solar irradiance variation between 1874 and 1988. *Science, 247,* 556–558.

Hansen, J. E. and A. A. Lacis, 1990. Sun and dust versus greenhouse gases: An assessment of their relative roles in global climate change. *Nature, 346,* 713–719.

Labitzke, J. and H. van Loon, 1990. Associations between the 11-year solar cycle, the quasi-biennial oscillation and the atmosphere: A summary of recent work. *Philosophical Transactions of the Royal Society of London, 330,* 577–589.

Schatten, K. H. and J. A. Orosz, 1990. A model for solar constant secular changes. *Geophysical Research Letters, 15,* 179–184.

Friis-Christensen, E. and K. Lassen, 1991. Length of the solar cycle: An indicator of solar activity closely associated with climate. *Science, 254,* 698–700.

Kuhn, J. R. and K. G. Libbrecht, 1991. Nonfacular solar luminosity variations. *Astrophysical Journal, 381,* L35-L37.

Lean, J., 1991. Variations in the Sun's radiative output. *Reviews of Geophysics and Space Physics, 29,* 505–535.

Mitchell, W. E. and W. C. Livingston, 1991. Line-blanketing in the irradiance spectrum of the sun from maximum to minimum of the solar cycle. *Astrophysical Journal, 372,* 336–348.

Reid, G. C., 1991. Solar total irradiance variations and the global sea surface temperature record. *Journal of Geophysical Research, 96,* 2835–2844.

Willson, R. C. and H. S. Hudson, 1991. The sun's luminosity over a complete solar cycle. *Nature, 351,* 42–44.

Hoyt, D. V., H. L. Kyle, J. R. Hickey, and R. H. Maschhoff, 1992. The Nimbus-7 solar total irradiance: A new algorithm for its derivation. *Journal of Geophysical Research, 97A* 51–63.

Kelly, P. M. and T. M. L. Wigley, 1992. Solar cycle length, greenhouse forcing, and global climate. *Nature, 360,* 328–330.

Lean, J., O. R. White, A. Skumanich, and W. C. Livingston, 1992. Estimating the total solar irradiance during the Maunder Minimum. *Eos, 73,* 244.

Schlesinger, M. E. and N. Ramankutty, 1992. Implications for global warming of intercycle solar irradiance variations. *Nature, 360,* 330–333.

Currie, R. G., 1993. Luni-solar 18.6 and solar cycle 10–11 year signals in USA air temperature records. *International Journal of Climatology, 13,* 31–50.

Books on Sun/Weather/Climate Relationships

Hahn, F. G., 1877. *Über die Beziehungen der Sonnenfleckenperiode zu meteorologischen Erscheinungen.* Ph.D. Thesis, University of Leipzig.

Secchi, A., 1877. *Le Soleil,* vols. 1 and 2. Gurthier-Villars, Paris.

Fritz, H., 1878. *Die Beziehungen der Sonnenflecke zu den magnetischen und meteorologischen Erscheinungen der Erde.* Haarlem, The Netherlands.

Egeson, C., 1889. *Egeson's Weather System of Sun-spot Casuality.* Turner and Henderson, Sydney, 63 pp.

Hutchins, D. E., 1889. *Cycles of Drought and Good Seasons in South Africa.* Times Office, Wynberg, 136 pp.

Fritz, H., 1893. *Die wichtigsten periodischen Erscheinungen der Meteorologie und Kosmologie.* Ph.D. Thesis, University of Leipzig.

Young, C. A., 1895. *The Sun.* D. Appleton, New York.

Lockyer, N., 1897. *The Sun's Place in Nature,* Macmillan, New York, 360 pp.

Hann, J., 1901. *Lehrbuch der Meteorologie.* Leipzig.

Hann, J., 1908. *Handbuch der Klimatologie,* 3 vols., Engelhorn, Stuttgart, 764 pp.

Abbot, C. G., F. E. Fowle, and L. B. Aldrich, 1908–1954. *Annals of the Astrophysical Observatory of the Smithsonian Institutions,* vols. 2–7. U. S. Government Printing Office, Washington, D.C.

Krümmel, O., 1911. *Handbuch der Ozeanographie,* part II.

Murray, J. and J. Hjort, 1912. *The Depths of the Ocean.* Macmillan, London, 821 pp.

Huntington, E., 1914. *The Climatic Factor as Illustrated in Arid America.* Carnegie Institution of Washington, Washington, D.C, 341 pp.

Bigelow, F. H., 1915. *A Meteorological Treatise on the Circulation and Radiation in the Atmospheres of the Earth and of the Sun.* Wiley, New York, 431 pp.

Douglass, A. E., 1919. *Climatic Cycles and Tree Growth.* Publication 289, Part 1. Carnegie Institution of Washington, Washington, D.C.

Helland-Hansen, B. and F. Nansen, 1920. *Temperature Variations in the North Atlantic Ocean and the Atmosphere: Introductory Studies on the Cause of Climatological Variations.* Smithsonian Miscellaneous Collections, 70, no. 4, 408 pp.

Huntington, E. and S. S. Visher, 1922. *Climatic Changes: Their Nature and Causes.* Yale University Press, New Haven, 329 pp.

Clayton, H. H., 1923. *World Weather, Including a Discussion of the Influence of Variations of Solar Radiation on the Weather and of the Meteorology of the Sun.* Macmillan, New York, 393 pp.

Huntington, E., 1923. *Earth and Sun: An Hypothesis of Weather and Sunspots.* Yale University Press, New Haven, 296 pp.

International Council of Scientific Unions, 1926. *First Report of the Commission Appointed to Further the Study of Solar and Terrestrial Relationships.* Paris, 202 pp.

Douglass, A. E., 1928. *Climatic Cycles and Tree Growth.* Publication 289, Part 2. Carnegie Institution of Washington, Washington, D.C.

Shaw, Sir N., 1928. *Manual of Meteorology.* vol. 2, *Comparative Meteorology.* University Press, Cambridge, England.

Glock, W. S., 1933. *Tree-ring Analysis on Douglass System.* Geological Publishing, Des Moines, 144 pp.

Abbot, C. G., 1934. *The Sun and the Welfare of Man.* Smithsonian Institution Series, New York, 322 pp.

Douglass, A. E., 1936. *Climatic Cycles and Tree Growth.* Publication 289, Part 3. Carnegie Institution of Washington, Washington, D.C.

Clayton, H. H., 1942–1943. *Solar Relations to Weather,* 2 vols. Clayton Weather Service, Canton, Mass. Vol. 1, 105 pp.; Vol. 2, 439 pp.

Huntington, E., 1945. *Mainsprings of Civilization.* Wiley, New York, 660 pp.

Johnson, M. O., 1946. *Correlation of Cycles in Weather, Solar Activity, Geomagnetic Values, and Planetary Configurations.* Phillips and Van Orden, San Francisco, 149 pp.

Peterson, W. F., 1947. *Man, Weather, Sun.* Charles C. Thomas, Springfield.

Stetson, H. T., 1947. *Sunspots and their Effects. From a Human Point of View.* McGraw-Hill, New York, 201 pp.

Johnson, M. O., 1950. *Cycles in Weather and Solar Activity.* Paradise of the Pacific Press, Honolulu, 224 pp.

Abbot, C. G., 1958. *Adventures in the World of Science.* Public Affairs Press, Washington, D.C., 150 pp.

Nupen, W. and M. Kageorge, 1958. *Bibliography on Solar-Weather Relationships.* American Meteorological Society, Washington, D.C., 248 pp.

Ellison, M. A., 1959. *The Sun and Its Influence. An Introduction to the Study of Solar-Terrestrial Relations.* Routledge and Kegan Paul, London, 237 pp.

Anonymous, 1961. *Solar Variations, Climatic Change, and Related Geophysical Problems.* Annals of the New York Academy of Sciences, 95 740 pp.

Eigenson, M. S., 1963. *Sun, Weather, and Climate* [in Russian], Gidrometizdat, Leningrad.

Keith, L. B., 1963. *Wildlife's Ten-Year Cycle,* University of Wisconsin Press, Madison, 201 pp.

Ellison, M. A., 1968. *The Sun and Its Influence. An Introduction to the Study of Solar-terrestrial Relations.* Elsevier Publishing, New York, 340 pp.

Fritts, H. C., 1976. *Tree Rings and Climate.* Academic Press, London, 567 pp.

White, O. R. (ed.), 1977. *The Solar Output and Its Variation.* Colorado Associated University Press, Boulder, 526 pp.

Herman, J. R. and Goldberg, R. A., 1978. *Sun, Weather, and Climate.* NASA SP-426, GPO, Washington, D.C., 360 pp.; reprinted 1985, Dover Publications, New York.

McCormac, B. M. and Seliga, T. A., 1979. *Solar-Terrestrial Influences on Weather and Climate.* D. Reidel Publishing, Dordrecht, 346 pp.

Pepin, R. O., J. A. Eddy, and R. B. Merrill (eds.), 1980. *The Ancient Sun. Fossil Record in the Earth, Moon, and Meteorites.* Pergamon, New York, 581 pp.

CNES, 1981. *Conference Proceedings, Sun and Climate.* Centre National d'Etudes Spatials (CNES), Toulouse, 480 pp.

Zastavenko, L. G., 1981. *Physical Principles of the Variability of Present-Day Climate.* Izdatel'stvo Nauka, Moscow, 160 pp.

Frazier, K., 1982. *Our Turbulent Sun.* Prentice-Hall, Englewood Cliffs, 198 pp.

McCormac, B. M. (ed.), 1983. *Weather and Climate Responses to Solar Variations.* Colorado Associated University Press, Boulder, 626 pp.

Schove, D. J., 1983. *Sunspot Cycles.* Hutchinson Ross Publishing, Stroudsburg, Pennsylvania, 410 pp.

Barchiia, S. P., 1985. *Astronomical Climate in the USSR.* Izdatel'stvo Nauka, Moscow, 176 pp.

Pecker, J. C. and S. K. Runcorn (eds.), 1990. *The Earth's Climate and Variability of the Sun over Recent Millennia.* Cambridge University Press, Cambridge, 692 pp.

Schatten, K. H. and A. Arking (eds.), 1990. *Climate Impact of Solar Variability.* NASA CP-3086, Washington, D.C. 376 pp.

Nesmes-Ribes, E. (ed.), 1995. *The Solar Engine and Its Influence on Terrestrial Atmosphere and Climate.* Springer-Verlag, Berlin, 561 pp.

Pap, J., C. Frohlich, H. S. Hudson, and W. K. Tobiska (eds.), 1995. *The Sun as a Variable Star: Solar and Stellar Irradiance Variations.* Kluwer Academic Publishers, Dordrecht, 319 pp.

Index